Titles in Series:

Why It's OK to Want to Be Rich
Jason Brennan

Why It's OK to Be of Two Minds
Jennifer Church

Why It's OK to Ignore Politics
Christopher Freiman

Why It's OK to Make Bad Choices
William Glod

Why It's OK to Enjoy the Work of Immoral Artists
Mary Beth Willard

Why It's OK to Speak Your Mind
Hrishikesh Joshi

Why It's OK to Be a Slacker
Alison Suen

Why It's OK to Eat Meat
Dan C. Shahar

Why It's OK to Love Bad Movies
Matthew Strohl

Why It's OK to Not Be Monogamous
Justin L. Clardy

Why It's OK to Trust Science
Keith M. Parsons

Why It's OK to Be a Sports Fan
Alfred Archer and Jake Wojtowicz

Selected Forthcoming Titles:

Why It's Ok Not to Think for Yourself
Justin Tosi and Brandon Warmke

Why It's OK to Be Fat
Rekha Nath

Why It's OK to Be a Socialist
Christine Sypnowich

Why It's OK to Be a Moral Failure
Robert B. Talisse

For further information about this series, please visit: www.routledge.com/Why-Its-OK/book-series/WIOK

Why It's OK to Trust Science

Why trust science? Why should science have more authority than "other ways of knowing?" Is science merely a social construct? Or even worse: a tool of oppression? This book boldly takes on these and other explosive questions—lodged by ideologues on the left and the right—and offers readers a well researched defense of science and a polemic addressed to its detractors.

Why It's OK to Trust Science critically examines the recent history of critiques of science, including those in academia from scholars like Bruno Latour, Simon Schaffer, and Thomas Kuhn. It then presents case studies drawn from recent advances in the field of dinosaur paleontology, showing how science generates objective knowledge, even during revolutionary episodes. The book next looks at how that same objective knowledge can be gained even when researching extremely complex issues, using climate science to distinguish between genuine skepticism – upon which science depends–from dogmatic denial.

The book is for anyone who needs thoughtful, razor sharp responses to the detractors of science—whether they be anti-vaxxers, climate change deniers, profit-seeking businessmen, or published relativists in the knowledge-making industries.

Keith M. Parsons is Professor of Philosophy and Humanities, Emeritus at The University of Houston-Clear Lake. His books include *God and the Burden of Proof* (1989), *Drawing Out Leviathan: Dinosaurs and the Science Wars* (2001), and *Bombing the Marshall Islands: A Cold War Tragedy* (2017; coauthored with Robert A. Zaballa).

Why It's OK: The Ethics and Aesthetics of How We Live

ABOUT THE SERIES:

Philosophers often build cogent arguments for unpopular positions. Recent examples include cases against marriage and pregnancy, for treating animals as our equals, and dismissing some popular art as aesthetically inferior. What philosophers have done less often is to offer compelling arguments for widespread and established human behavior, such as getting married, having children, eating animals, and going to the movies. But if one role for philosophy is to help us reflect on our lives and build sound justifications for our beliefs and actions, it seems odd that philosophers would neglect arguments for the lifestyles most people—including many philosophers—actually lead. Unfortunately, philosophers' inattention to normalcy has meant that the ways of life that define our modern societies have gone largely without defense, even as whole literatures have emerged to condemn them.

Why It's OK: The Ethics and Aesthetics of How We Live seeks to remedy that. It's a series of books that provides accessible, sound, and often new and creative arguments for widespread ethical and aesthetic values. Made up of short volumes that assume no previous knowledge of philosophy from the reader, the series recognizes that philosophy is just as important for understanding what we already believe as it is for criticizing the status quo. The series isn't meant to make us complacent about what we value; rather, it helps and challenges us to think more deeply about the values that give our daily lives meaning.

KEITH PARSONS

Why It's OK
to Trust Science

Routledge
Taylor & Francis Group
NEW YORK AND LONDON

Designed cover credit: Andy Goodman. © Taylor & Francis

First published 2024
by Routledge
605 Third Avenue, New York, NY 10158

and by Routledge
4 Park Square, Milton Park, Abingdon, Oxon, OX14 4RN

Routledge is an imprint of the Taylor & Francis Group, an informa business

© 2024 Taylor & Francis

The right of Keith Parsons to be identified as author of this work has been asserted in accordance with sections 77 and 78 of the Copyright, Designs and Patents Act 1988.

All rights reserved. No part of this book may be reprinted or reproduced or utilised in any form or by any electronic, mechanical, or other means, now known or hereafter invented, including photocopying and recording, or in any information storage or retrieval system, without permission in writing from the publishers.

Trademark notice: Product or corporate names may be trademarks or registered trademarks, and are used only for identification and explanation without intent to infringe.

Library of Congress Cataloging-in-Publication Data
A catalog record for this title has been requested

ISBN: 978-0-367-61640-3 (hbk)
ISBN: 978-0-367-61641-0 (pbk)
ISBN: 978-1-003-10581-7 (ebk)

DOI: 10.4324/9781003105817

Typeset in Joanna and Din
by Apex CoVantage, LLC

Access the Support Material: www.routledge.com/9780367616410

Contents

Acknowledgments — ix

Introduction: What Has Science Done for Me Lately? — 1

1. The "Science Wars" and Why They Had to Be Won — 21

2. The Facts About Social Construction — 43

3. Thomas Kuhn: Foe of Science? — 61

4. Thomas Kuhn: Friend of Science? — 85

5. Can We Have Good Science and the Right Values? — 103

6. Dinosaur Revolutions — 139

7. How We Know About Big, Complex Things — 171

Conclusion: What Is *Really* Wrong with Science — 203

Notes — 215
Bibliography — 221
Index — 229

Acknowledgments

I would like to thank Andrew Beck of Routledge for suggesting this project and supporting it throughout its production. Three anonymous reviewers made many valuable suggestions that greatly enhanced the project. Professor Andrew E. Dessler of Texas A & M University kindly read a draft of my chapter on climate change and made a number of valuable suggestions. Any errors remaining are, of course, mine alone. Professor Heather Douglas of Michigan State University graciously sent me a copy of her 2016 Descartes Lectures, *The Rightful Place of Science: Science, Democracy, and Values*, delivered at Tilburg University in the Netherlands.

Introduction
What Has Science Done for Me Lately?

We take science for granted. Every day we do things and use devices that would have so astounded people a few centuries ago that they would only marvel at them or perhaps fear them as some sort of black magic. Familiarity, they say, breeds contempt. I think that it is more accurate to say that it breeds indifference. I do not mean that people are indifferent to those quintessential products of science, our personal electronics. On the contrary, they are objects of obsession. I mean that people are indifferent to the brilliant science and engineering that went into those ubiquitous devices. A book-length exposition of the history and pre-history of the smartphone would encompass a large chunk of the accumulated scientific knowledge of our culture. How many people would read such a book? Not nearly as many as would want to get the app for the latest game.

It is probably even more accurate to say that familiarity breeds a sense of entitlement. It is human nature to feel that we are owed benefits we have long enjoyed. We only think about those benefits when they are denied. Having enjoyed the benefits of science and technology for generations, we take these for granted and only think about scientists and engineers when they fail to give us what we want or even slightly irritate us. If a software upgrade temporarily inconveniences us, we complain loudly. In the nineteenth century, new scientific theories and discoveries were front-page news. Now, discoveries that should astound us are given much less attention than the latest doings of royalty. Such is our insouciance about science that, if prompted, we would probably ask, "What has science done for me lately?"

To remind ourselves of what science has meant for our culture and for our lives, let's recall what life was like and how people understood things a mere 500 years ago. In the history of the Earth, or even the history of humanity, 500 years is just yesterday. If we figure three generations per century, then count back 15 generations from yourself and you are there. Yet, there is an enormous gulf between what people believed and how they lived then and what they know (or could know) and how we live today. Let's focus on the Western world since modern science, which is now genuinely international, had its beginnings in the West.

Five hundred years ago news traveled as fast as the swiftest horse or sailing ship. Things had hardly improved since 490 BCE when the famous runner Philippides gasped with his dying breath the news of the Athenians' victory at Marathon. In general, travel was slow and dangerous, not to mention uncomfortable, even for the rich and powerful. Speaking of comfort, though the Romans had invented central heating, human dwellings, from palaces to hovels, relied on fireplaces. Air conditioning was centuries away. Daily tasks such as cooking, cleaning, and laundry were very laborious and time-consuming. That is why those who could afford them hired servants to do the drudgery. Cooking was from scratch over a fire or on a wood stove. Even in my grandmothers' time an entire day, usually Monday, had to be set aside as wash day as clothes had to be boiled in a cauldron, hand-scrubbed with washboards, hand-rinsed (with water drawn from a well), and pinned to clotheslines to dry.

Five centuries ago, things were much worse when it came to health and well-being rather than mere convenience and comfort. In a world with no antibiotics, any injury could lead to amputation or death as there was a real danger of infection. An animal bite was doubly worrisome since it not only posed the usual dangers of wounds but also the terrifying possibility of rabies. Infectious diseases ran rampant, even among royalty. As late as 1861, Queen Victoria's husband, Prince Albert, died of typhoid, sending the Queen into a state of permanent mourning. Childbirth was dangerous and if a woman lived to have eight children (of course, there were no effective contraceptives), she would be lucky to have all of them survive to adulthood. A visit to an old graveyard will show you the toll childhood diseases took on families.

Doctors were often helpless in dealing with common maladies, and, due to sheer ignorance, often compounded the problems and spread disease. Basic concepts of sanitation and hygiene were unknown. Cities were fetid as chamber pots were dumped into the street. Before antisepsis and anesthetics, surgery was an option only for the truly desperate. The causes of infectious diseases were unknown. The learned doctors of the Sorbonne attributed the Black Death (bubonic plague) of the fourteenth century to a malignant conjunction of planets. Five hundred years ago medical schools still taught the theories of Galen, the eminent physician of the Roman Empire. People knew little about the internal organization of their own bodies, though this was about to change with the publication in 1543 of Andreas Vesalius' great work of anatomy *De humani corporis fabrica* (On the Structure of the Human Body).

Five centuries ago people lived with food insecurity. They were always one failed harvest away from famine. Even under the best of circumstances, the agricultural practices of the day yielded far less per acre than is produced by scientific farming techniques. If agriculture had not advanced since then, only a small fraction of today's nearly eight billion human beings could be fed. In short, without scientific medicine and agriculture, YOU, gentle reader, probably would not be here.

Astronomy was the most advanced science five hundred years ago. In astronomy, as in all other sciences, Aristotle (384–322 BCE) was the recognized authority. Aristotle maintained that the universe was a series of concentric crystalline spheres, centered upon the earth, and each sphere bore one of the heavenly bodies. The heavens were perfect and changeless. Meteors and comets were atmospheric phenomena, and the atmosphere extended to the sphere of the Moon. The Moon and all heavenly bodies were perfect spheres that moved in perfect circles at a constant rate. They were not constituted of the four elements that comprise earthly things—earth, water, air, and fire— but were made of the luminous *aether*. The Earth was known to be a sphere and, just 30 years before, it had been found to contain an entire New World. Yet, the best estimate of the age of the Earth was still the biblical one of just a few thousand years. Fossils were known, but they were not recognized as the remains of once-living creatures.

Further, as historian David Wootton observes, even the most learned people believed things now recognized as superstition or pseudoscience. They believed in astrology and that base metals can be turned into gold. They were sure that witches and demons tormented the world and employed their malevolent powers to cause many misfortunes. Further:

> He [the learned man of that time] believes that a murdered body will bleed in the presence of the murderer. He believes that there is an ointment which, if rubbed on a

dagger which has caused a wound, will cure the wound. He believes that the shape, color, and texture of a plant can be a clue to how it will work as medicine because God has designed nature to be interpreted by mankind . . . He believes that the rainbow is a sign from God and that comets portend evil. He believes that dreams predict the future, if we know how to interpret them.

(Wootton 2015, pp. 6–7)

I mention these facts not at all to disparage the people of five hundred years ago or to accuse them of gullibility or foolishness. People were just as smart then as they are today, and they based such beliefs on the best information available at the time. Guided by their lights, past scientists made the most reasonable judgments they could, and that is all we can require of anyone.

Here I am simply pointing to the fact that is obvious—or should be obvious—that modern science has vastly expanded our knowledge of the world, and, even more importantly, has given us the intellectual tools to achieve that understanding and surpass it. We may not be any smarter than we were 500 years ago, but our science is. Further, technology has changed our lives in innumerable ways, and perhaps now is even changing human nature itself.

"Ah," the critic will reply, "but has not science produced nightmares as well as wonders? What about nuclear weapons, nerve agents, and weaponized pathogens produced in biological warfare laboratories? What about the noxious by-products of science and technology, such as air and water pollution? In climate change are we not facing a global crisis of our own making? Has not science made warfare far more destructive and horrific?" These are good questions.

Consider war. War has always been a grim, gruesome, horrible business. Homer gave us detailed accounts of just where the spear went into the chest and the agonies of the dying. Ancient and medieval warfare was hand-to-hand, up close, and personal. You could smell the breath of the man you killed. Modern warfare makes violent death less personal but no less gruesome. Perhaps the most famous poem in English from the Second World War was Randall Jarrell's "The Death of the Ball Turret Gunner." The ball gun turret hung from the belly of the B-17 and B-24 bombers to defend against attacks from below. The attacks were conducted by skilled Luftwaffe pilots flying the deadly Messerschmitt 109 and Focke-Wulf 190 fighters. These fighters were armed with cannon and heavy machine guns. The effect of such weapons on the human body was atrocious:

> From my mother's sleep I fell into the State,
> And I hunched in its belly till my wet fur froze.
> Six miles from earth, loosed from its dream of life.
> I woke to black flak and the nightmare fighters.
> When I died they washed me out of the turret with a hose.

Or consider nuclear weapons. Nuclear weapons are a product of beautiful physics and of engineering of the highest order. In the Manhattan Project, the concentrated genius of many of the world's most powerful intellects produced a weapon of unprecedented power. That power was first unleashed in the pre-dawn darkness of the New Mexico desert on July 16, 1945, in the Trinity Test of the plutonium implosion design. Prior to the test, all the scientists had were figures on blackboards. Some thought that it would fizzle. It did not fizzle. When the countdown reached zero at 5:29 A.M., and the equivalent of 21,000 tons of TNT detonated, the desert was filled with an unearthly light, "brighter than a thousand suns," a spectacle one observer, physicist Kenneth Bainbridge, called "a foul and awesome display," recognizing the implications for the future of humanity, Bainbridge said of the bomb's creators, "Now we are all sons of bitches."

Science therefore seems a terribly mixed blessing. Science gives knowledge and knowledge is power. We are ambivalent about power. As Lord Acton reminded us, power corrupts. On the other hand, we celebrate the empowerment of the previously marginalized and powerless. Power enhances our ability to do good or ill, and it is a necessary condition for getting anything done at all. Further, if power is repudiated, it does not go away; it just goes to someone else. Science, then, is a mixed blessing because it gives us the power to achieve great benefit or inflict great harm.

There are no unmixed blessings. Religion is a mixed blessing. Democracy is a mixed blessing. Wealth is a decidedly mixed blessing. That said, most people regard religion (theirs, anyway), democracy, and money as good—very good—things. Science certainly deserves admiration if any of those things do. On the other hand, our admiration for science need not be unqualified. We can regard science with the sort of respect that we accord to any powerful and potentially dangerous thing. Such respect means that we must impose ethical restrictions upon science, limiting what we *can* do by what we *should* do. Scientific power, like political power, must be constrained, guided, and balanced.

Yet science should not be judged purely in terms of its material products, but chiefly with respect to its intellectual and, indeed, spiritual benefits. The three greatest things I learned in school were the extent of the cosmos, the age of the earth, and the world of the very small.

The vastness of the universe is literally incomprehensible. You can speak of millions or billions of light years and even calculate in those terms, but the sense of scale that we have from our ordinary experiences is totally inadequate to grasp astronomical dimensions. On a clear fall evening, far away from the light pollution that ruins the night sky for urban and suburban dwellers, a person with good eyesight can see the Andromeda Galaxy, 2.5 million light years away. The light that is reaching your eyes started on its journey long before our species existed, yet the Andromeda Galaxy is a member of what astronomers call the "local group" of galaxies. Using instruments such as the Hubble Telescope, we can see galaxies that existed long before the formation of our solar system. What is truly remarkable is that we not only can *see* things at that distance, but we can *understand* them. The laws of physics were as good ten billion years ago and ten billion light years away as they are here and now.

Like so many nerdy kids, my first intellectual interest was dinosaurs. A lost world of huge, fantastic beasts, some longer than two city buses, fired my imagination and sparked my wonder. The discovery of deep time in the late eighteenth century profoundly affected its discoverers. One reported a feeling of giddiness, as if standing upon a precipice and peering down into the abyss of time. Truly one of the great intellectual revolutions was the realization that the time scale based on biblical chronology was short by six orders of magnitude.

Equally marvelous to me was the world of atoms and molecules. It is still a breathtaking thought to realize that there is a world within the world, a realm of the very small that is as multifarious and intricate as the world of our ordinary experience. Further, we know that everything that happens, from the merest flicker of an eyelid to a supernova involves changes in that microworld. So basic is the knowledge of the very small to our body of scientific knowledge that Richard Feynman, perhaps, after Einstein, the leading physicist of the twentieth century, said that if a catastrophe destroyed all of science except for a single sentence, the sentence that should be saved would tell us that everything is made of tiny particles called "atoms."

So, science provides us with the deep satisfaction of understanding, but it also gives us rewards that are properly termed "spiritual," that is, science can evoke a deep sense of reverence, awe, and wonder. The depth of feeling that science can induce is akin to—I would say identical with—the sense of religious awe that is felt in the presence of the sacred. Einstein referred to the universe as *Der Herrgott*, The Lord God. The beautiful and majestic peroration that ends Darwin's *Origin of Species* discloses Darwin's love of living things and his admiration for the terrible yet marvelous process of evolution that has generated them. Aristotle said that in even the humblest of organisms there is something of the divine. There is a silly idea, often promulgated by romanticists and poets (even some of the great ones), that scientific knowledge somehow destroys our sense of wonder and ruins the beauty of natural things. On the contrary, the more we understand the world, the deeper, richer, and more fulfilling will be our appreciation of it.

Whatever problems, therefore, may come with science—and we will see in the conclusion that science needs to do a lot of housecleaning—science is better in every respect than any alternative. We are immeasurably richer—and not just materially—because of science. It is an enterprise fully worthy of our deepest appreciation. Yet science has its enemies. Some would debunk it or dethrone it from its position of respect and explain its success as due to politics, privilege, rhetoric, and chicanery. Other more insidious enemies are happy to pay lip service to science, but really work to undermine it because science threatens their profits or dogmas. This book will therefore be an unabashed apologetic for science and a polemic addressed to its detractors.

This has been the justification for the book, and we now turn to its plan.

Events of worldwide significance sometimes begin as academic debates. Disputes between professors can play out on the world stage and impact millions of lives. For instance, with some justification, critics of the French Revolution such as Edmund Burke blamed the philosophers of the Enlightenment for the destruction of the *ancien régime*. Of course, large historical events are complexly caused, but there is no question that ideas have consequences. That is why it is wrong to dismiss academic disputes as tempests in teapots. Ideas that will have great impact for good or ill may start in the ivory tower, but they cannot be immured there.

The first chapter, "The Science Wars and Why They Had to Be Won," reviews one of the nastier academic rows. In the 1990s, disagreements over the nature, credentials, and status of science grew so strident and rancorous that the dispute came to be called "the science wars." The conflict climaxed in 1996 when physicist Alan Sokal submitted an article to the journal *Social Text*, a leading venue for academic critics of science. The article, "Transgressing the Boundaries," was intentionally loaded with nonsense and gibberish. When *Social Text* published the article, Sokal revealed the hoax and those stung were furious. Beneath the brouhaha, serious issues were at stake concerning critiques made by radical feminists, sociologists of knowledge, and postmodernist theorists. The upshot of these critiques was that science is a social construct, an excrescence of politics, white male privilege, rhetoric, intimidation, and subterfuge. Defenders of science replied with vigorous, often trenchant arguments.

Why resurrect a controversy from a quarter century ago? Time adds perspective, so that we can see which arguments were important and which were more peripheral. Really, though, academic dust-ups generally settle nothing. On the contrary, issues are hashed and re-hashed, combatants succumb to mutual exhaustion, and few minds are changed. In fact, the conflict smolders on, and though the debates now may be less strident, the disagreements still run deep. In his bestseller *Enlightenment Now*, Steven Pinker notes the continued academic disparagement of science. Historian David Wootton, quoted previously, felt the need to write his massive *The Invention of Science*, published in 2015, to set the record straight about the Scientific Revolution, a record that he holds continues to be distorted by relativism and social constructivism among historians of science. The issues from the "science wars" remain relevant and interesting, so it seems like a good time for a retrospective.

At the heart of much science critique is philosophical relativism, the claim that truth is not absolute, but relative to culture, theory, or conceptual scheme. Supporters of relativism draw upon two arguments associated with the philosopher W.V.O. Quine, the "holism thesis" and the "underdetermination thesis." The former says that any theory can be made compatible with any evidence by appropriate adjustments in our other beliefs. The second says that any number of theories are compatible with any body of evidence. Relativists draw upon these theses to claim that no theories can claim to be simply true, but true only relative to perspective. The first chapter considers these arguments and rebuts them, arguing that the holism and underdetermination theses do not adequately support conceptual relativism.

Perhaps the most vocal science warriors were the social constructivists, those who argue that the claims of science lack objective support and that the methods of science are not reliable guides to truth or rationality. Instead they claim that science is a cultural artifact and nothing more; its conclusions and methods are determined by politics, rhetoric, power plays, and various historical, sociological, and psychological factors. The second chapter examines the claims of prominent constructivists Bruno Latour, Steven Shapin, and Simon Schaffer, and concludes that constructivist arguments are both historically and philosophically deficient.

The third and fourth chapter focus on the works of Thomas Kuhn. Kuhn's most famous book is *The Structure of Scientific Revolutions*, first published in 1962. Why devote two chapters to a work that was published six decades ago and has been extensively discussed? Academic skepticism about science can largely be traced back to Kuhn's *Structure*. Since Kuhn's work chronologically preceded the arguments examined in the first chapter, why not put the chapters on Kuhn first? The reason is that the science critics of the "wars," to a man or woman, referred back to Kuhn as justifying and inspiring their critiques. It is therefore logical to move from the various more recent critiques back to the common source that was claimed to motivate them.

Kuhn certainly did not see himself as an enemy of science, and in his later career strongly opposed what he regarded as the misappropriation of his ideas by relativists and social constructivists. Yet, there is no question that in *Structure* Kuhn said things that are plausibly interpreted as undermining important elements of the rationality of science. In particular, he denied that observations are a neutral arbiter between opposing theories and said that theories determine their own data. Further, though theories might use the same terms, those terms undergo a radical variance in meaning between the competing theories, rendering those theories "incommensurable." Finally, Kuhn said that scientists undergo a sort of "conversion," akin to religious conversion, when they change theories. In a sense, their whole world changes.

Within a decade of the appearance of Kuhn's book, philosophers had published a number of astringent responses to Kuhn, rebutting those striking claims about scientific rationality. By the time Kuhn sought to clarify and qualify his distinctive claims, denying that he was defending relativism or irrationality, it was too late. By that time many other scholars, including historians, sociologists of knowledge, and literary theorists had picked up on Kuhn's radical strains and were running with the ball.

They still are. Steven Pinker reports that Structure is still the second most assigned book on science in universities (Pinker 2018: 395). Kuhn is very much a living presence in any discussions of the rationality of science, and a book like this one must still come to terms with him. Also, in all fairness, Kuhn got some things exactly right. Since the book remains widely read and influential, and since its ideas are still important and controversial, we need to make clear just what Kuhn got right and what he, or at least his more radical readers, got badly wrong. To try to sort out where Kuhn really stood with respect to scientific rationality, the third chapter will be titled "Thomas Kuhn: Foe of Science?" and the fourth will be "Thomas Kuhn: Friend of Science?"

Perhaps science is most controversial when it clashes, or appears to clash, with our moral, social, or political values. The fifth chapter asks, "Can We Have Good Science and the Right Values?" The results of science cannot be expected to conform to the demands of ideologies, sets of prior beliefs that serve as creeds for various groups. The attempt to force science to serve doctrinal demands corrupts science and reduces it to just another ideology. This chapter begins by examining just such an attempt by the feminist writer Sandra Harding to adapt science to the service of her values. Most of the chapter considers the much more creditable efforts of Heather Douglas to understand just how moral and social values can be used by scientists without corrupting scientific objectivity.

In Chapter 6 "Dinosaur Revolutions," I will focus on the very big changes in dinosaur paleontology over the past few decades. These changes would certainly seem to deserve to be classified as "revolutions," since dinosaurs have been re-envisioned in radically new ways. "Dinosaur" used to be a metaphor for obsolescence, but now we see that these creatures were marvelously adapted to dominate their environment. They were agile, active, intelligent, and quite possibly some were warm-blooded. We now know that birds did not just descend from the dinosaurs, but *are* modified theropod dinosaurs, and, conversely, the dinosaurs of the Mesozoic were far more bird-like than lizard-like.

Perhaps the most remarkable change in our understanding of dinosaurs was how the (non-avian) dinosaurs became extinct. According to the "impact" theory the mass extinction that ended the 160 million year reign of the dinosaurs—and numerous other creatures—was the result of the massive impact of an extraterrestrial body, an asteroid or comet. To say that the response of traditional paleontologists to this theory was skeptical would be a very significant understatement; the debate was often bitter and even abusive. Now, however, it is generally, but not universally, accepted among paleontologists that a large impact played at least a very significant role in the end-Cretaceous mass extinctions.

The sixth chapter will ask whether these very significant changes of outlook involved anything like Kuhnian incommensurability or extreme discontinuity. In paryticular, I look at one major participant in these debates, leading paleontologist David Raup, and ask whether he underwent anything like a Kuhnian "conversion" in changing his mind. More basically, the chapter asks whether dinosaurs should be considered social constructs or whether we can be confident that we do have some very well-confirmed theories about these marvelous denizens of deep time. The upshot is that science, as exemplified by the case of dinosaur paleontology, can undergo radical changes by conservative means. That is, the boring old story about science—that its changes are driven by discovery, data crunching, and the application of rigorous methods—is more accurate than the exciting stories proposed by Kuhn and the social constructivists.

The seventh chapter is titled "How We Know about Big, Complex Things." We may make a distinction, not wholly satisfactory, between "global" science skeptics and "local" ones. Global skeptics are those, like some of the more extreme academic science critics of the "science wars," who argue that science, in its results and methods, is wholly a social construct, determined in structure and content by politics or social influences. "Local" skeptics often announce their loyalty to science, but then direct skepticism at particular scientific claims such as evolution or human-caused climate change. This is not really a satisfactory distinction since those who deny established scientific claims when those claims conflict with their dogmas or vested interests are hypocrites who only pay lip service to scientific methods and ideals, respecting them only when they are convenient.

The sciences repudiated by the "local" skeptics are sciences that deal with complex phenomena, and which are supported by complex and multifaceted evidence. These sciences of complexity, such as evolution and climatology, do not depend upon one or a few pieces of evidence. The evidence supporting evolutionary theory, or anthropogenic climate change is copious, and cumulative. Such theories are supported as the best explanations of large bodies of very diverse evidence. One thing that makes an explanation "best" is that it achieves what philosophers of science call "consilience," that is, it unites seemingly disparate and independent phenomena by subsuming them under a common explanatory framework. Thus, evolution by natural selection provided a theoretical basis for incorporating and understanding data from anatomy, embryology, geographical distribution, paleontology, taxonomy, and other fields. As for climate change, no one result or discovery clinched the case. Rather, evidence built as alternatives were excluded, critics were addressed, error bars narrowed, and stricter tests were applied.

In the early stages of a scientific controversy, when hypotheses are not rigorously constrained by evidence, many different claims are plausible, and controversy will be rife among the experts. However, as the amount and quality of evidence accumulates, one hypothesis will emerge as the best-confirmed, and alternative hypotheses will be excluded. Eventually, consensus emerges among the qualified parties. Skepticism is rational up to a point, but not after that. How is ambiguity decreased and convergence to certainty achieved?

I will look at climate science and see how, from tentative beginnings, the claim that human activity is changing the global climate has become established beyond a reasonable doubt. The chapter will also make a distinction between the kind of skepticism necessary for science and the fake "skepticism" of those who will not be convinced by any amount of evidence.

The conclusion, "What Is *Really* Wrong with Science," will consider how science is all too often its own worst enemy. As the present book argues, those external enemies of science—those motivated by ideology and avarice—fail to score any points against science. However, in 2020, psychologist Stuart Ritchie published an important and disturbing book titled *Science Fictions: How Fraud, Bias, Negligence, and Hype Undermine the Search for Truth*. Ritchie shows that scientists themselves have done more to undermine public trust than all the ideologues and obscurantists. Scientists work under intense duress as they compete for grants, jobs, and status. The high-pressure and hypercompetitive milieu was imposed with the aim of rewarding only the best science and excluding the bad. Ironically, the effect has been the opposite, perversely incentivizing bad science.

Ritchie offers several suggestions for removing the perverse incentives that afflict science and replacing them with incentives for greater transparency, rigor, and honesty. I will argue more generally that the malaise that has degraded science parallels a broader malady that has afflicted all of academe, and not just the sciences. Hostile politicians, abetted by bottom-line administrators, have sought, with considerable success, to impose a "business model" on universities, whereby education is a product, students are customers, and professors (or scientists) are tools. When education—or science—becomes a product, then moving product becomes the goal, and this means that success is measured in terms of such quantities as the number of semester credit hours taught and the number of studies published.

I will therefore conclude by arguing for two broader cultural changes in the administration of our academic/scientific enterprises, and in our attitudes towards scientific achievement. First, I argue for the end of what historian Jerry Z. Muller calls "the tyranny of metrics." Today's cult of metrics assumes that all aspects of human performance, including academic and scientific performance, can be quantified, and numerically assessed so that the rewards can be apportioned on the basis of supposedly objective measures. The imposition of metrics was intended to make administrative decisions transparent and objective, but, in fact, the metrics obsession distorts, oversimplifies, and corrupts our assessments of human achievement. I therefore argue that metrics should not replace, but complement judgments based upon what Aristotle called the "practical wisdom" derived from human experience. Second, I argue that the intense individualism of scientific status and credit be replaced by a much more communal pride. Science is preeminently a shared and cooperative enterprise. The basic unit of scientific achievement is the relevant community, not the brilliant individual. Science is something WE do; it is an interdependent activity through-and-through. Science is the glory of our species, and we should ALL take pride in its achievements.

Three appendices are posted in the online accompaniment of this book. The first is a summary of the first 13 sections of Kuhn's *Structure of Scientific Revolutions*. The second is a response to Richard Rorty's essay "The World Well Lost" in which I argue that the world is not well lost. The third is a list of 21 facts supportive of anthropogenic climate change. I consider these appendices as organic parts of the book rather than ancillary to it. However, space limitations required that they be put online.

GLOSSARY

Certain terms occur repeatedly and centrally in this book. Before proceeding, let me define them here.

Postmodernism: This is a hard term to define precisely. Broadly, postmodernism was a movement in the humanities and the arts that flourished in the 1980s and 1990s. It could mean different things to different people. To its supporters, it was a movement that was playful, eclectic, anti-hierarchical, and puckishly disrespectiul of all canons, traditions, and purportedly universal standards. From this perspective, postmodernists looked like the heirs of the Dadaists who painted mustaches on the Mona Lisa and mounted urinals in art museums to ridicule "high" culture. To its detractors, postmodernists were not merry pranksters but purveyors of fashionable nonsense, pseudo-intellectual obscurantists who dispensed impenetrable jargon and advocated a kind of nihilistic relativism that abolished rationality and meaning.

Relativism: According to Simon Blackburn in his Oxford Dictionary of Philosophy, relativism is defined as:

> The permanently tempting doctrine that in some areas at least, truth itself is relative to the standpoint of the judging subject . . . . Relativism may be a global doctrine about all knowledge, or a local doctrine about some area (aesthetics, ethics, or judgments of secondary qualities, for example). The aspects of the subjects supposed to determine what truth is "for them" may include historical,

cultural, social, linguistic, or psychological background, or brute sensory constitution.

(Blackburn, 326)

Relativism can also be about justification, that is, the standards and norms invoked to substantiate truth claims. Thus, a relativist with respect to justification will concede that experimental results may be good evidence for truth claims within the conceptual framework of physical science, but deny that it should take universal priority over, say, the claim within a religious framework that some beliefs are justified by revelation.

Social Construct: A social construct is anything created by the collective decision of the relevant community. A mountain is not a social construct. It is an objective, external fact. Whether we *call* a topographic feature a "mountain" or maybe just a "hill" depends on criteria that are socially constructed. Vertical displacements of the landscape obviously vary continuously in size, from the merest molehill to Mt. Everest. Geographers can agree on some arbitrary height, say 1000 meters above the average elevation of the surrounding countryside, and stipulate that any discrete geological structure of that height or greater be designated as a "mountain." That criterion is a social construct. The Matterhorn is not a social construct. It is an objective reality that existed long before it was seen and named by humans.

The "Science Wars" and Why They Had to Be Won

1

The Sokal Hoax

In the fall of 1994 physicist Alan Sokal submitted an article to the prominent academic periodical *Social Text*. It was unusual for a physicist, or any scientist, to submit an article to this venue. *Social Text* had achieved its prominence by publishing articles by leading scholars of the "academic left," including such luminaries as Edward Said and Cornel West. Though the country had swerved to the right, with sweeping wins for Republicans in the 1994 mid-term elections, the academic left flourished. Feminists and postmodernists published copiously, as did the practitioners of the burgeoning fields of cultural studies and science studies. One historically anomalous feature of much of this activity was its seeming hostility towards science.

Since the Enlightenment, progressive politics had traditionally allied itself firmly with science. However, the leftist academics of 30 years ago had acquired a reputation as harsh critics of science, characterizing it as patriarchal, racist, oppressive, exploitative, and spurious in its claims to objectivity and rationality. These critiques prompted an acidulous response titled *Higher Superstition: The Academic Left and Its Quarrels with Science* (1994) coauthored by biologist Paul R. Gross and mathematician Norman Levitt. In a famous essay written in the 1950s, scientist and novelist C.P. Snow had famously decried the rift between the "two cultures"—the scientific and the literary. By the mid-1990s that rift had become a Grand Canyon sized gap.

Yet here was a submission by a respected theoretical physicist to the publishing hub of academic left activism. Sokal's Article was titled "Transgressing the Boundaries: Towards a Transformative Hermeneutics of Quantum Gravity" (Sokal 2000: 11–45). The very wording of this title, with phrasing invoking the transgressing of boundaries and hints of chic postmodernist theorizing ("transformative hermeneutics"), seemed designed to resonate with the editors of *Social Text*. Surely, they must have thought, this was an encouraging sign; at least one physicist had seen the light! However, Sokal's article was a sham—a farrago of trendy jargon, puns, mathematical and scientific howlers, non-sequiturs, and groundless assertions. Normally, of course, one would expect that any such article would be quickly weeded out by the rigorous vetting of editors or peer reviewers. Yet Sokal's send-up was published by *Social Text*—as a serious article—in the spring/summer issue of 1996. Sokal immediately revealed the hoax, to the intense consternation of *Social Text* and its supporters, and to the delight of their critics. The sting apparently revealed the laziness, ignorance, incompetence, and perhaps even the intellectual dishonesty of the left-wing science critics.

The news of Sokal's hoax went global, drawing comment from sources that seldom reported on academic controversies. The academic left rallied around *Social Text* and replied to Sokal in highest dudgeon. *The New York Times* published "Professor Sokal's Bad Joke" by an outraged Stanley Fish. Fish is an eminent literary theorist and at the time was the director of Duke University Press, publisher of *Social Text*. Fish regarded Sokal's "joke" as motivated by two mistakes: (1) He misunderstands what it means for something to be a "social construct," equating "constructed" with "not real," and (2) he wrongly thinks that the sociology of science, the discipline that studies the social factors that influence the scientific process, as an attempt to debunk science.

To Sokal, the left-wing science critics seemed determined to collapse the fact/construct distinction so that, say, DNA is reduced to a figment of scientists' collective imagination rather than a real molecule. Fish replies that to say that something is a social construct is not to deny that it is real (Fish 2000: 82). He uses the example of baseball. Balls and strikes are social constructs, defined by the rules of baseball. Yet they are real enough that large sums of money are paid to those who can produce a maximum of strikes and a minimum of balls. Science, of course, is concerned with things that are not social constructs, but, says Fish, the practice of science, with its variable and evolving methods, norms, and judgments, is just as much a congeries of social practices as baseball, and just as apt a subject for sociological investigation (Fish, 2000: 83). Sokal's other mistake, he says, is to think that those engaged in science studies are equating science with baseball, that is, reducing science to an esoteric game and denying that it deals with objective, external realities. By making this mistake, says Fish, Sokal erroneously regards the sociology of knowledge as a competitor to science, when, in fact, it is a parallel and independent discipline and does not attempt to offer a competing account of the physical world (Fish, 2000: 83).

By making these mistakes, says Fish, Sokal was motivated to play his malicious prank. The upshot is that it is Sokal who undermines intellectual standards (Fish 2000: 83). Fundamental to any academic enterprise is the basic trust between colleagues that reports and scholarship are legitimate and not spiteful hoaxes.

A very different take on the hoax was offered by Paul Boghossian, professor of philosophy at New York University.

Writing in the *Times Literary Supplement*, Boghossian titled his essay "What the Sokal Hoax Ought to Teach Us" (Boghossian 2000: 172–182). Boghossian picks up on Fish's baseball example and notes that, while, of course, the rules and terms of baseball are socially constructed, once they are defined, the umpire is charged with making a judgment (notoriously fallible) of an objective *physical* fact, i.e. the position of a baseball within or outside of a defined strike zone. Likewise, says Boghossian, though theories are, of course, human constructs, and though the concepts and vocabularies whereby theories are framed no doubt reflect the contingencies of human capacities and limitations, this is no reason to think that those theories cannot express objective facts about the physical world (Boghossian 2000: 180). Terms such as "electron," "gene," and "tectonic plate" had to be devised and we can tell the story about how they came to be, but there is no reason to think that these terms do not pick out and refer to objective physical things. "Gene" may be a social construct, but genes are not.

As Boghossian sees it, the editors and contributors to *Social Text*, whom he broadly terms "postmodernists," are in fact seeking to debunk science and demote it from its status as the privileged, authorized way of investigating the world. He takes them as making the characteristic claim of relativism, namely, that there are "many ways of knowing" and that Western science is only one of many equally valid such "ways" (Boghossian 2000: 177). He cites some professional archaeologists who say that the conclusions of archaeology are no more authoritative than the claims of Native American mythology (Boghossian 2000: 177). If archaeology says that the first humans entered North America by crossing from Asia across the Beringia land bridge and Native American myth says that their ancestors emerged onto the continent from a subterranean spirit world, then postmodernist relativism requires that both claims be regarded as equally "true."

Boghossian argues that relativism, whether with respect to truth or justification, is incoherent and self-defeating (Boghossian 2000: 178–179). Relativism about truth entails that a claim and its opposite can both be true if there exists some perspective relative to which each claim can be true. However, in this case, relativism itself can be true for relativists, and its opposite—that truth is objective and not relative—can be true for those who hold that truth is objective. Yet the whole aim of aggressive postmodernists is to definitively debunk the notion of the objectivity of truth. Apparently, you can only debunk objective truth by assuming objective truth! With respect to relativism about justification, it is true that a claim and its negation might be equally justified, as, for instance, when there is little evidence either way. However, relativism about justification entails that *any* claim can be justified by adducing its own rules of evidence. In that case, relativists must admit that their claims about the relativity of justification can be no more justified than the claims of those who deny the relativity of justification. Thus, Boghossian argues, relativism quickly sinks into a morass of self-refutation.

As for the editors of *Social Text*, Boghossian holds that they cannot escape blame (Boghossian 2000: 174–175). Sokal's intentional errors were so blatant that Boghossian concludes that there are only two possible explanations for the fact that the hoax was published: (1) The editors were so scientifically and mathematically illiterate that they failed to notice the arrant nonsense. (2) They recognized the errors but did not care. All that mattered was that the piece supported their ideology. Neither option reflects well on the editors.

PHONY WARS?

So, the battle lines were drawn, and the stakes could not be higher—the future of science and scientific rationality in our intellectual culture. Or was it really such a big deal? Richard Rorty, one of the best-known American philosophers of the twentieth century, wrote a deflationary article for *The Atlantic Monthly* titled "Phony Science Wars" (Rorty 1999). Rorty's essay is, in part, a review of philosopher of science Ian Hacking's *The Social Construction of What?* What makes the science wars "phony" for Rorty is the characterization of the dispute by those who see it as a conflict between good guys who defend science, truth, and rationality, and the bad guys—postmodernist and social constructivist obscurantists who want to dissolve the solid achievements of science into a vapor of fashionable ideology.

For Rorty, the debate really comes down to a conflict of intuitions that are nearly as old as Western philosophy itself:

> These alternating intuitions have been in play ever since Protagoras said "Man is the measure of all things" and Plato rejoined that the measure must instead be something nonhuman, unchanging, and capitalized — something like The Good, or The Will of God, or The Intrinsic Nature of Physical Reality. Scientists who, like Steven Weinberg, have no doubt that reality has an eternal, unchanging, intrinsic structure which natural science will eventually discover are the heirs of Plato. Philosophers like Kuhn, Latour, and Hacking think that Protagoras had a point, and that the argument is not yet over.
>
> (Rorty 1999)

So, as Rorty sees it, the debate is not a confrontation between rationality and irrationality, but a continuation of a clash between two venerable intellectual traditions, each of which is worthy of respect. One side has the "intuition" that there is an external world with intrinsic, determinate properties existing independently of our descriptions, and the other side intuits that "reality" is determined by our percepts and concepts (I consider Rorty's arguments for losing "the world" in Appendix II). Rorty therefore holds that the so-called science wars are nothing new and that its impassioned combatants should cool the rhetoric and recognize that they are just the latest participants in an ancient and respectable debate. They should focus on providing serious answers to serious questions rather than hoaxing, harrumphing, and waxing self-righteously indignant.

To anyone who is familiar with the views of Rorty, such sermonizing must appear more than a bit hypocritical. Rorty called himself a "pragmatist," thereby locating himself in the American pragmatist tradition along with William James, Charles Sanders Peirce, and John Dewey. Actually, his most famous publication, *Philosophy and the Mirror of Nature* (Rorty 1979), is rightly regarded as a postmodernist manifesto. In this work, Rorty defends a thoroughgoing relativism that he calls "epistemological behaviorism." He claims that the idea that knowledge can have solid foundations had been exploded by the work of philosophers such as Dewey, Martin Heidegger, Ludwig Wittgenstein, and W.V.O. Quine. The only alternative is an epistemological variant of "When in Rome, do as the Romans do." That is, as he puts it, say what your society allows you to say; justify your beliefs by appealing to the rules of evidence contingently sanctioned by your intellectual community. You cannot do any better than that.

It follows that no such set of standards can claim superiority over others; archaeology cannot appeal to rules of evidence that are superior to those employed by Native American mythologists. Scientific rationality is indeed just "another way of knowing" and should not be regarded as having authority over other, equally valid, cognitive enterprises. Rorty's hope is that by dethroning certain types of discourse and delegitimizing their hegemonic claims, we can all just relax, end the obsessive search for The Truth, and indulge in a free-flowing and mutually edifying "conversation of mankind." The essence of Rorty's critique can be summarized with the words of the old reggae song, "Don't worry. Be happy."

However, anyone who *does* think that there is something special about science, and that it *really is* a superior way to discover objective truth about the natural world, will rightly regard Rorty as an enemy of scientific rationality. To such persons, Rorty's *Atlantic* article, while posing as detached from the science wars and denouncing them as "phony," really is just another shot fired by one of the combatants.

The issues of the science wars were not phony but real and important. On the one side were those whom we might term "scientific rationalists." We may identify scientific rationalism with four core theses:

1. The Reality Thesis: There is a physical world with intrinsic, determinate properties that exist independently of human concepts, descriptions, or worldviews.
2. The Accessibility Thesis: That physical world is at least partially accessible to us, that is, by observation (directly or with instruments), experiment, and measurement we may interact with that world and ascertain certain facts about it.
3. The Warrant Thesis: Those facts sufficiently constrain our theorizing to warrant acceptance of some theories and hypotheses as the most rational, probable, or credible accounts we possess.

4. The Rationality Thesis: Scientific communities reach consensus chiefly on the basis of objective, unbiased evidence and rational argument, even during revolutionary episodes.

The first thesis merely asserts, *contra* Rorty, that there is a real world with an intrinsic nature. The second says that scientific methods, practices, and techniques give us access to that world and permit us to know facts about it. The third thesis asserts the objectivity of warrant, that is, that certain facts *really do* authorize some claims about the world but not others—no matter what society "lets us say." For instance, the fossils *really do* substantiate some claims about dinosaurs and not others. The fourth point asserts that, *pace* social constructivists, scientists frequently are convinced by reasons rather than consensus being driven by politics or ideology.

Scientific rationalists also hold that science has succeeded in establishing a large body of discoveries, that is, truths about the natural world. Each of these truths may be held tentatively, and ongoing research will no doubt lead to revisions, but it is absurd to think that the whole corpus of scientific belief could be wrong. Scientific rationalists do not deny that politics and social factors have influenced the course of science. They maintain, however, that scientific methods can—and frequently do—give the last word to nature. Further, they see science as progressive in at least two senses: (1) Over time, science discovers more and more about the intrinsic nature of the universe, and (2) science regularly discovers better ways of doing science, that is, methods and techniques that, *inter alia*, provide more data, more varied data, more relevant data, clearer data with more signal and less noise, or more precise data allowing for more rigorous testing of hypotheses.

The science critics of the academic left who opposed scientific rationalism were a multifarious group and offered a diversity of perspectives and arguments. However, we may identify three prominent theses offered against scientific rationalism:

1. **Relativism**: The claim that truth and/or justification is not absolute but relative to perspective, conceptual framework, or culture. Science seeks to establish "the truth" about the natural world, but there are many equally valid ways of knowing, each with its own standards and theories fully authorized within that worldview.
2. **Social constructivism**: The claim that science, both in its results and in its methods, is a social construct through and through. Scientific "facts," e.g., about DNA, dinosaurs, quarks, or galaxies, are generated by the social and political interactions of scientists, particularly by rhetoric, power plays, privilege, and jockeying for prestige and grants. Scientific methods likewise are not neutral and reliable means of accessing nature but are adopted because they are the practices of socially sanctioned and politically privileged groups.
3. **Political rectitude**: The claim that the scientific ideal of a neutral, disinterested objectivity is spurious and that, in fact, the values, standards, and practice of science are thoroughly political. All science is political science, and unavoidably so. In that case, since science must be based upon politics, it is essential that it be based upon the right politics. Science should be based upon politics that liberate and empower the marginalized rather than privilege the already overprivileged.

I will address the first of these theses in this chapter, the second in the next, and the third in the fifth chapter.

The issues between scientific rationalists and their critics were real and important—not phony. It was therefore important for rationalists to win the science wars. I think they did.

RELATIVISM AND THE UNDERDETERMINATION ARGUMENT

I believe that Boghossian is right that relativism is the philosophical heart of many of the critiques of science that issued from the academic left during the science wars. Indeed, it is the a priori assumption built into social constructivist accounts of science (Wootton 2015: 44). We therefore need to look more closely at relativism and one of the main arguments supporting it. Can we offer a sweeping, knockdown refutation of relativism of the sort that Boghossian proposed? Some relativists are aware of the self-refutation trap and attempt to turn the tables. What if you could confute the defenders of objective truth by showing them that *by their own* assumptions, the idea of objective truth is untenable? The more sophisticated relativists attempt to do this. One of the most frequently invoked arguments for relativism was not proposed by a postmodernist *poseur*, but is associated with the name of one of the hardest-nosed of analytic philosophers, W.V.O. Quine (1908–2000).[1]

Quine referred to an argument of Pierre Duhem (1861–1916), French physicist and historian of science, on the relationship between evidence and hypotheses. Duhem noted that the idea that there can be a crucial experiment that selectively refutes just one particular hypothesis overlooks a vital point about testing. Hypotheses are never tested in isolation, but only in conjunction with indefinitely many auxiliary hypotheses. Experimental tests must make many assumptions about the functionality of the equipment, the adequacy of the experimental design, the absence of interfering factors, and the interpretation of results. When a theoretical prediction is made, and the prediction is not confirmed by experiment or observation, then *something* is wrong somewhere, but just where? It *may* be that a particular hypothesis is wrong, but perhaps the reason the prediction did not occur is due instead to a problem with one or more auxiliary hypotheses. Perhaps the observation was in error or was misconstrued. Perhaps (as often happens) equipment was being employed at the extreme limit of its capabilities, and results are disputable. Or perhaps something totally unanticipated has skewed the outcome. Deciding where to put the blame for predictive failure is seldom straightforward.

The upshot is that the old saying about scientific method that "Man proposes; nature disposes" is simplistic. By refusing to honor our predictions, nature shows us that *something* is wrong with our ideas, but we must figure out exactly what. So, Duhem correctly stated the complexity of experimental tests of hypotheses. However, as appropriated by Quine, the argument was construed as having a much more radical implication, which we may call the "holism thesis" (HT). The HT asserts that when we encounter evidence contrary to our predictions, that contrary evidence challenges not just a particular hypothesis, or even the whole theory within which that hypothesis is embedded, but the *whole* of science or even the *whole* of our beliefs. For Quine, the consequence of holism is that we can always preserve a favored theory from refutation by *any* evidence. All we have to do when contrary evidence arises is to make sufficient adjustments to other parts of our system of beliefs. Even our logical and mathematical principles can be modified if we have sufficient motivation to do so. The big question, of course, is whether it is *rational* to save a hypothesis at any cost. More on this below.

An even more radical thesis appears to follow from the HT, which we may call the "underdetermination thesis" (UT). The HT implies that diehard proponents of an old theory can always hold on to that theory by making appropriate adjustments to the theory so that apparently contrary evidence can be accommodated. A well-known example from the eighteenth century was the adjustments made to phlogiston theory in the face of contrary results indicated by very careful experiments made by Antoine Lavoisier. Phlogiston theory proposed that in combustion, the burning material releases a substance called phlogiston. If burning releases such a substance, then the products of combustion should weigh less than the material prior to burning. However, Lavoisier's experiments indicated that the product of combustion weighed more, and it was this result that led to the realization that combustion is an oxidation reaction. However, phlogiston theorists accommodated this result by proposing that phlogiston has negative weight, and so the remnants of combustion will weigh more when no longer buoyed up by the released phlogiston.

Whether one regards such an answer as an ingenious idea or a disgracefully *ad hoc* ruse is irrelevant to the logical point that phlogiston theory can be altered to be made consistent with Lavoisier's results. This means that both phlogiston theory and oxidation theory, though incompatible with each other, can be made compatible with (i.e., not contradicted by) the experimental results. An extrapolation of this reasoning is that two competing theories can be made compatible with *any possible* experimental results. A further extrapolation takes us to the claim that *indefinitely many* competing theories can be made compatible with *any possible* evidence. This last claim is the basis for the "underdetermination thesis" (UT). The UT is the claim that all theories are radically underdetermined by their evidence, even all possible evidence. That is, no matter how much evidence we have, we will never find just one theory that is consistent with that evidence. On the contrary, it will always be possible to devise indefinitely many other competing theories that can be made consistent with the evidence, even if we have all *possible* evidence. Therefore, say defenders of the UT, no amount of evidence for a theory can pick it out from among its indefinitely numerous competitors and certify it as the single true one.

In principle, it seems to be right that alternative and incompatible theoretical schemes can always be invented to accommodate any set of phenomena, just as infinitely many curves can be drawn to fit any finite set of points. It follows that no finite set of data points can uniquely *determine* (i.e., entail) a given theory. Epistemological relativists have seized on the UT to argue that therefore an endless multiplicity of possible theories is compatible with any amount of evidence. Radical relativism about theories is assumed to follow. (See the discussion of Shapin and Schaffer in the next chapter.) In the UT relativists have perhaps found an argument that appeals to no premise that a believer in objective truth could deny, and so they escape self-refutation and instead attribute it to their opponents.

In the actual practice of science, is underdetermination a real problem? Are scientists overwhelmed by a superabundance of possible theories? Not at all. On the contrary, just coming up with *one* good theory is often very hard. It is interesting to ask why the UT seems like such a deep and serious problem for (some) philosophers but not for scientists. How do scientists pare down a potential infinity of theories to just one or maybe a few rival theories? Do their practices obviate the philosophical problem?

Most obviously, scientific methods are designed to severely restrict the "wiggle room" in the interpretation of experimental results and focus upon specific hypotheses as the reason for the failure or success of prediction. Neutralizing alternative explanations is why control groups are so important. If group A is, in all relevant respects, just like group B, except that no one in group A smokes cigarettes and everyone in group B smokes two packs a day, then significantly increased morbidity and mortality in group B is reasonably attributed to group B's smoking habits. Of course, controlling for all relevant factors is not easy, and that is why multiple independent studies are called upon to confirm such conclusions.

Further, and equally importantly, theories must jump through many hoops before they even become acceptable *candidates* for testing *vis-à-vis* the evidence. These pre-test criteria impose much stronger requirement on theories than mere compatibility with the evidence. Logical compatibility with the data is a necessary condition for a candidate theory, but it is far from a sufficient one. As Friedel Weinert notes, each theoretical model exists within a "constraint space," a set of empirical and theoretical factors that severely limit the types of theories a scientific community will consider as worthy of testing (Weinert 2009: 61–62). In any particular case, the conjunction of these constraints can suffice to rule out whole sets, indeed infinite sets, of alternative models. In actual scientific practice, these constraints serve to narrow the field to a few viable models for testing

We may identify three categories of pre-test or candidacy requirements (These are not necessarily the only types of such requirements, just three prominent ones.):

1. **Regulative assumptions**. Like all rational inquiry, scientific inquiry must be guided by certain assumptions. An important kind of assumption is the "regulative" assumption. A regulative assumption frequently constitutes a general

prohibition against certain types of scenarios. Conservation laws are among the best-known instances of regulative assumptions, employed to make basic judgments about what can happen and what cannot. Conservation laws are not unfalsifiable, but many conservation principles are regarded as so fundamental that theories that blithely ignore or contradict them are not taken seriously as candidates for testing.

The fact that an inquiry rests on assumptions does not mean that its results must be dubious. Detectives and plumbers make assumptions, and crimes still get solved and leaks repaired. Assumptions are not arbitrary. They are made because long and often painful experience has shown that they are useful, even necessary guides to truth.

2. **Background knowledge**: All scientific knowledge is in principle tentative, but some things are regarded as known to a practical certainty. Such certainties are said to constitute our "background knowledge." For instance, basic physical constants such as c, the speed of light in a vacuum, and h, the Planck Constant are taken for granted. Some principles, such as the laws of thermodynamics, are so abundantly confirmed and so basic to our understanding of the universe that they may also be taken for granted. Any new theory that disregards the most solidly grounded of our background beliefs probably will not get a hearing.

3. **Theoretical virtues**: Good theories are expected to evince certain virtues, such as testability, simplicity, i.e., employing an economy of means to explain a wealth of data, mathematical rigor, consilience (subsuming diverse kinds of evidence under a unified theory), clarity, and coherence—relations of reciprocal support among the elements of a

theory. Cobbled-together or crackpot theories will generally be notably deficient in virtue, and so judged unworthy of testing.

In the actual practice of science, therefore, underdetermination is quite tractable. Scientists can invoke pre-test criteria to exclude whole classes of possible theories and leave only a few genuine candidates. Underdetermination is not a problem for scientists given their assumptions, practices, and standards.

Relativists might reply that these assumptions, practices, and standards are ones that other "ways of knowing" do not accept. True. Astrology does not study the sky the way that astronomy does. Christian fundamentalists do not approach the history of the earth as geologists do. Remember, though, that the UT was invoked to show that the objectivist assumptions built into scientific rationality were self-refuting. This means that relativists must address scientific *rationalists'* assumptions, not those of practitioners of other "ways."

Still, hardboiled proponents of the UT might argue that scientific assumptions, values, and practices, while perhaps pragmatic necessities for doing science, do not really get to the heart of the problem, which is an issue of logic and not pragmatics. Theories postulating, say, gods or witches will be dismissed by scientists as, for various reasons, unvirtuous or as contradicting regulative assumptions. However, that does not mean that god or witch theories cannot entail the evidence as much as any theory scientists would favor. In that case, god or witch theories will be as compatible with the evidence as scientific ones. Scientific preferences therefore leave the *logical* problem of UT untouched.

Larry Laudan does get to the heart of the issue by noting that to seriously challenge scientific rationalism and its claims of the objective confirmation of theories by evidence, a much stronger thesis of underdetermination is needed. Merely claiming that innumerable theories are *compatible* with the data or even *entail* the data is far too weak. What is needed is something like this:

> Any theory can be shown to be as well supported by any evidence as any of its known rivals.
>
> (Laudan 1999: 93)

A scientific rationalist might be quite willing to admit that two or more rival theories are compatible with the evidence—or even entail the evidence. The important question is whether the evidence equally supports two or more of these rivals. For instance, often defenders of a theory will favor it as the best explanation of the evidence. Two theories can equally entail the evidence, but one will be a much better explanation. To take a simple case, consider the scene from *Robinson Crusoe* where Crusoe had been alone on his island for many years. Then, one day, walking on the beach, he sees an impression in the shape of a human footprint. Consider two rival theories:

T_1: The impression was caused by a human foot.
T_2: The impression was caused by a purely accidental interaction of wind and tide that made a depression in the sand that looks exactly like a human footprint.

Both T_1 and T_2 entail the evidence, but, obviously, T_1 is far more plausible, and, of course, Crusoe immediately concludes that he is no longer alone on his island.

When you consider actual known rival theories, it is hardly ever the case that two such rivals are *exactly* equivalently related to the evidence. Hence, a heavy burden of proof must be borne by anyone who would say that *any* two rival theories will be equally supported by *any* evidence. To bear the burden of proof imposed by Laudan, the defender of UT would have to consider all the ways that evidence can support theory and then show that in no such way can evidence support one theory over its known rivals. It is hard to see how this could be done. Footprints really do seem to be much better evidence for feet than for vagaries of wind and tide. Frankly, it is just hard to take seriously the suggestion that all theories are equally supported by any evidence, e.g., that the theory that the Apollo moon landings were faked is evidentially equivalent to the theory that they really happened, or that the theory that ancient astronauts built the pyramids has the same evidential support as the theory that the Egyptians built them.

Yet cannot one *always* appeal to the HT to show that two or more theories are equally supported by the evidence? Not really. Again, not all alternatives entailing the data are equally credible; some are perverse. As philosopher Thomas Nagel amusingly notes, I could propose the theory that hot fudge sundaes every day will make me lose weight. I then explain away the higher readings on my bathroom scales by saying that the laws of mechanics governing the function of my scales are suspended whenever I step on them (quoted in Boghossian 2006: 127). Obviously, it would be pathologically irrational to make such an excuse.

In other words, sometimes it is just not reasonable to invoke the HT. For instance, an astronomer that made a telescopic observation that appeared to invalidate an accepted theory could blame the telescope rather than the theory. However, the idea that failure of the telescope is just as likely as failure of the theory is not reasonable:

> The idea, however, that in peering at the heavens through a telescope we are testing our theory of the telescope *just as much* as we are testing our astronomical views is absurd. The theory of the telescope has been established by numerous terrestrial experiments and fits in with an enormous number of other things we know about lenses, light, and mirrors. It is simply not plausible that, in coming across an unexpected observation of the heavens, a rational response might be to revise what we know about telescopes.
>
> (Boghossian 2006: 128; emphasis in original)

The general point of the Nagel and Boghossian examples is this: In many cases it is just not rational to save a theory from potentially falsifying evidence by rejecting one or more of the auxiliary hypotheses. Sometimes the cost of saving a hypothesis is just too high. The practice of saving a sacrosanct hypothesis at *any* cost is a characteristic stratagem of the invincibly ignorant. For instance, if you are determined at all costs to protect a literal reading of the creation story in the Book of Genesis, how do you deal with the fact that there are fossil creatures, such as dinosaurs, that appear to have flourished and gone extinct eons before the events recounted in Genesis? Were dinosaurs on the Ark with Noah? "Yes!" say fundamentalists. There is a "creation science" museum in Kentucky that has full-size displays of Noah's Ark with dinosaurs on board. So, you can insulate any theory from falsifying evidence, but sometimes at the cost of being ridiculous.

Underdetermination arguments therefore do not pose an insuperable problem for those who hold that science can arrive at objective truth.[2]

To do justice to the issue of relativism, a whole book would be needed, not just part of one chapter. (For a patient, careful, but ultimately critical study of relativism see Kirk 1999.) Let me lay my cards on the table: I think that the intellectual case for relativism is weak. In my view, the popularity of relativism is not due to the strength of its arguments, but rather to its presumed moral merits. History has shown the dire consequences of inquisitions, whether religious or secular, pursued by those who believed themselves in possession of absolute truth. As intellectual historian John Herman Randall notes, the means employed by inquisitors may have been questionable, but their logic was impeccable (Randall: 1976: 79). Those who think that they have possession of the one absolute and supremely important truth must regard all schismatics, dissidents, and heretics as criminals—worse than murderers said, St. Thomas Aquinas. Further, racism, colonialism, and imperialism were justified by the allegedly superior morals, law, and religion that it was the "white man's burden" to impose upon

. . . fluttered folk and wild,
Your new-caught sullen peoples,
Half devil and half child.

If Absolute Truth can justify such enormities, then the solution would seem to be a relativism that takes all voices to be equal in the "conversation of mankind."

Yet, the great champions of liberation and justice have never been relativists, but those who themselves possessed what they thought were transcendent truths. "We hold these truths to be self-evident, that all men are created equal, and are endowed by their Creator with certain unalienable rights." Frederick Douglass was hardly a relativist about slavery, and neither was Martin Luther King, Jr. about segregation. Gandhi did not hold that British rule was bad only relative to the views of the Indian National Congress. No one has ever stood on a barricade waving a flag emblazoned with "It's true for me." Relativism is a set of tepid talking points, entertaining conversation for the faculty club, but never a great liberating vision. Relativism appears to support diversity, multiculturalism, and equality, but can just as easily be employed as a cynical defense of dishonesty. When called out for mendacity, the Trump Administration spokesperson said that their lies were "alternative facts." Thus, the postmodernist chickens came home to roost in the Trump White House.

The Facts About Social Construction

2

Science is unquestionably influenced by social factors, sometimes amusingly. For 45 years, from 1934 to 1979, the Carnegie Museum in Pittsburgh displayed its prize *Apatosaurus* skeleton with the wrong head. It appears that social factors, such as the desire to have an impressive specimen to display to the public, rather than scientific considerations motivated the decision to attach the head (Parsons 1997). Sometimes the social influence on science is insidious. In 1981 Harvard paleontologist Stephen Jay Gould published *The Mismeasure of Man* (Gould 1981). This work documented how nineteenth-century anthropologists, using what they thought were objective measures, concluded that Western Europeans had the largest brains and therefore must be smarter than anybody else. Some German researchers even claimed to find that German skulls were more capacious than the skulls of other Europeans. Clearly, racist and ethnocentric biases had skewed the science, and such examples can be multiplied *ad nauseam*.

What if all science is like this, an excrescence of politics, rhetoric, ambition, privilege, bias, and a stew of other such unsavory factors? What if science is a social construct through and through, and inevitably and irremediably so? Consider DNA. In 1953, James Watson and Francis Crick published the double-helix model of the DNA molecule. In 1968, Watson published *The Double Helix*, his own account of

DOI: 10.4324/9781003105817-3

the discovery that won the Nobel Prize for himself and Crick (Watson 2012). The book was sort of a tell-all exposé that included many statements that a later age would find jarring and sexist. Watson's book reminds us that scientists are human and subject to all the frailties of human nature. In scientific communities, as in all communities, there are politics, ambition, hierarchy, feuds, and bias. Surely, though, no informed person would say that there was no discovery of the double helix and that the vast edifice of molecular biology built upon that discovery must be spurious.

Yet in 1979 Princeton University Press published *Laboratory Life: The Construction of Scientific Facts* by Bruno Latour and Steve Woolgar. In this book, Latour and Woolgar argue that scientific facts—or maybe we should say "facts"—are indeed the product of the social interactions of scientists (Latour and Woolgar 1979). What starts as a hypothesis is transformed into a fact when its advocates succeed in manipulating other scientists into agreement. The manipulation is performed by rhetoric, negotiation, power politics, wheeling and dealing, grandstanding, *ad hominem* attacks, and intimidation. Nature has nothing to do with it, or, rather, "nature" is the result of scientific warfare where the winners write the textbooks.

How do we understand such a project? Latour and Woolgar base their conclusions on an empirical study of the behavior of scientists at the Salk Institute, a well-known research facility in La Jolla, California. Yet if all "facts" are constructed, then what about the alleged facts uncovered by the research of Latour and Woolgar? Their project appears self-refuting; in the attempt to debunk science they appear to have debunked themselves. It is tempting to dismiss their claims out of hand. Yet all authors, even those whose conclusions are highly antithetical, should be read with some degree of charity, so let's try to get clearer on the claims and arguments.

Latour is not the most lucid of writers, but he states his case in a focused and explicit way in a portion of his 1987 book *Science in Action* (Latour 1987). At one point he discusses the infamous case of Prosper Blondlot and N-rays. In 1895, Wilhelm Roentgen discovered X-rays to much acclaim, including the Nobel Prize. In 1903, at the University of Nancy in France, Prosper Blondlot, a reputable physicist, claimed to have found a new kind of ray emitted from stressed metal. He called them "N-rays," and published a number of reports announcing his supposed discovery. Other physicists were skeptical because they could not reproduce Blondlot's results. Eventually, an American physicist, Robert W. Wood, visited Blondlot's lab to see for himself how the supposed N- rays were detected. At one point, while Blondlot was busy with his equipment and was reporting the observation of N-rays, Wood quietly removed an essential piece of the experimental apparatus. This should have instantly stopped the production of the supposed rays, but Blondlot continued to claim to detect them. For Wood, and soon for the entire physics community, it became clear that there were no N-rays and that Blondlot was a victim of wishful thinking and the unfortunate human capacity for "seeing" what is not there. Isaac Asimov pronounces the usual judgment:

> There seems no question but that Blondlot was utterly sincere. Nevertheless, the N-rays were an illusion, his reports proved worthless, and his scientific career was blasted.
> (Asimov 1993: vol. III, 216)

So, there never were any N-rays and Blondlot was just deluded. Latour strongly objects to this account:

> It would be easy enough for scientists to say that Blondlot failed because there was "nothing really behind his

N-rays" to support his claim. This way of analyzing the past . . . crowns the winners, calling them the best and the brightest and . . . says that the losers like Blondlot lost simply *because* they were wrong . . . Nature herself discriminates between the bad guys and the good guys. But is it possible to use this as a reason why in Paris, in London, in the United States, people slowly turned N-rays into an artefact? Of course not, since at that time today's physics obviously could not be used as the touchstone, or more exactly since today's state is, in part, the *consequence* of settling many controversies such as the N-rays.

(Latour 1987: 100, emphasis in original)

It is illegitimate to say that Blondlot failed to spot N-rays simply because they never existed. On the contrary, there was no fact of the matter about N-rays until the physics community had settled the matter to its satisfaction. Then, and only then, could it be said that, in fact, there were no N-rays. Prior to the collective decision of the physics community to construct the "facts" about the non-existence of N-rays, there was no fact of the matter. N-rays neither existed nor failed to exist. Likewise, Blondlot's "failure" to detect N-rays was a construct, determined by the collective decision of physicists to regard it as a failure. The outcome of that controversy—the non-existence of N-rays—cannot be used to explain Blondlot's failure to detect them. Latour makes this explicit as his "Third Rule of Method:"

> Since the settlement of a controversy is the *cause* of Nature's representation, not the consequence, we can never use the outcome—Nature—to explain how and why a controversy has been settled.
>
> (Latour 1987: 99; emphasis in original)

Well, why not? Why did people fail to see the *Titanic* dock in New York in 1912? Because it never arrived. It would be absurd to say that the news reports about the sinking of the *Titanic* were the reason it never docked. An iceberg sank the *Titanic*, not news reports, and the fact that people did not know about the iceberg until they heard the news is irrelevant. It would be similarly absurd to say that Blondlot's N-rays were not missing until Wood's fine job of debunking was communicated. Latour seems to be making the elementary mistake of conflating the fact of the matter with people's realization of the fact. Once the fact is made plain, whether it is the sinking of the *Titanic* by an iceberg or the nonexistence of N-rays, then *the fact itself* is made available to explain what happened.

Latour would no doubt regard such a robustly commonsensical reply as question-begging and as having missed his point. What, then, is the point? The point seems to be the one mentioned by Rorty, namely, that for people like Latour, reality has no intrinsic nature. There is no external world with determinate properties waiting upon our discovery. It is our descriptions, concepts, and categories that give a determinate shape to things and thereby make them "real." Rorty puts it like this:

> Take dinosaurs. Once you describe something as a dinosaur, its skin color and sex life are causally independent of your having described it. But before you describe [something] as a dinosaur, or as anything else, there is no sense to the claim that it is "out there" having properties.
> (quoted in Boghossian 2006: 27)

Here I simply have to confess that the claim that I find nonsensical is that reality does not have an intrinsic nature, that there is not *some* way that things are independently of our descriptions. The idea that there were no dinosaurs independently of our descriptions of them seems absurd, so absurd that, to adapt a quip from George Orwell, only a very educated person could be made to believe it.

Perhaps the most charitable way to take Latour is that he is making a claim not about reality but about what we can know. Reality never comes to us raw and naked, but only as clothed in descriptions and theories. Therefore, we have no access to a putative pristine nature, just as it is, but only as we represent it, and our representations are products of the scientific process. Consequently, we cannot appeal to nature itself as the reason that a scientific controversy turned out as it did since all we can know are our representations, and representations, by definition, are human constructs, determined by contingencies of time, place, and culture. Therefore, the idea that reality has an intrinsic nature is not discarded because it is false, but because it is *useless*.

It is difficult to see the supposed power of this line of thought-some version of which is offered by many thinkers of a constructivist or relativist hue. It is, of course, necessarily true, a tautology in fact, that you cannot describe something without using words or think about it without employing concepts. However, it does not follow from such trivial truths that the intrinsic nature of the physical world is ineffable or occult. From the fact that we represent the world with concepts it does not follow that we cannot know the world but only our representations (see Appendix II). The important question is not whether we can think about nature without thinking about it, but whether our representations are corrigible by interaction with the physical world so that less accurate representations can be replaced by better ones.

Consider dinosaurs again. At first there were only a few teeth and fragments of bone. Early paleontologists reconstructed dinosaurs by analogy with the creatures they knew, construing them, for instance, as scaled-up iguanas. As more fossil discoveries were made, the big-lizard interpretation did not hold up, and the uniqueness of these extraordinary creatures became apparent. In 1841 Richard Owen recognized dinosaurs as a distinct order. In the twentieth century, fabulous fossil finds in Asia and South America added immensely to the data. In recent decades, the tools of paleobiology have breathed life into the bones and authorized a far more comprehensive view of dinosaurs as living organisms (see Chapter 6). Dinosaurs were once conceived as hulking, obsolescent, nitwits, but are now seen as astonishingly successful creatures that dominated the terrestrial habitat for 160 million years.

So, we are led to the question of scientific methods. Do scientific methods provide objective evidence, trustworthy indications of truth about the physical world? Are there non-socially constructed facts about the justification of scientific claims? When people changed their minds about dinosaurs, did they do so for objective reasons? Are dinosaurs social constructs?

The year 1985 saw the publication of Steven Shapin and Simon Schaffer's *Leviathan and the Air-Pump* (Shapin and Schaffer 1985). Shapin and Schaffer resurrect an obscure seventeenth-century debate between the philosopher Thomas Hobbes, author of the classic of political philosophy *Leviathan*, and the scientist Robert Boyle and his allies. Boyle was an advocate of the experimental method, typified by his researches with the newly invented air-pump. The efficacy of experimental methods had been known since ancient times. Even the Bible contains a fine example of a crucial experiment, the contest at Mt. Carmel between Elijah and the prophets of Baal

(I Kings, 18). So, Boyle and the founders of modern science did not invent the experimental method; they just found out how to do it better, and moved it from the methodological margins to the center (Wootton 2015: 327).

Shapin and Schaffer depict Hobbes as rejecting the new-fangled methodology and insisting that rigorous deductive reasoning from indubitable first principles was sufficient for science. Of course, Boyle won the debate, and experimental methods have been central to the practice of physical science ever since. Shapin and Schaffer argue that Boyle won, and experimental methods have reigned ever since, not because experiment is actually a better way of doing science, but because Boyle and his allies were more socially connected and politically astute, playing the game better than Hobbes in the context of Restoration politics. A strike against Hobbes was that he was widely considered an atheist in an age that abhorred atheism. Boyle, on the other hand, was notably pious. Also, science at that time was being professionalized and part of that process was the adoption of distinctive scientific methodologies, the mastery of which would mark the professional.

In general, Shapin and Schaffer conclude, the methods of science are always a product of local social and political influences. Scientists think that their methods permit them to access objective evidence about the natural world, but actually they adopt methods that are socially conformist and politically expedient. Shapin and Schaffer therefore conclude their book by stating plainly that it is *we ourselves* that determine the content of science, and not the physical world:

> As we come to recognize the conventional and artifactual status of our forms of knowing, we put ourselves in a

position to realize that it is ourselves and not reality that is responsible for what we know.

(Shapin and Schaffer 1985: 344)

This last quote is constructivism in a nutshell. By adopting such a position, Shapin and Schaffer must face the problem of self-reference, as Cassandra Pinnick notes:

... the same arguments that would show ... that Hobbes and Boyle fixed their scientific beliefs because of political commitments and social alliances within Restoration society, would also show that Shapin and Schaffer's preferred analysis is merely a reflection of their (Shapin and Schaffer's) political commitments and social alliances—unless, of course, historians are immune, in ways that scientists are not, to social determination.

(Pinnick 1998: 229)

To expand on Pinnick's important point a bit, Shapin and Schaffer's causal account in terms of political and cultural factors must be superior to the rationalists' claim that scientists progressively discover better ways of doing science (i.e., new methods and techniques that improve our ability to discover and confirm truths about the natural world). To discredit the scientific rationalists' position, Shapin and Schaffer would have to argue one of the following: (a) there are no objectively better or worse methods for guiding empirical inquiry; (b) there may be reliable methods, but scientists cannot distinguish between bad, good, and better methods; or (c) there may be reliable methods, and scientists could, in principle, recognize them, but, in fact, scientists always adopt methods due to their political and social expedience.

If Shapin and Schaffer were to argue any of these theses, then they would appear to put themselves in a difficult position. If they claim (a), then they need to say why they have access to good methods but scientists do not. If they say (b), then they have to explain why they can recognize good methods when they see them but scientists cannot. If they say (c), then, as Pinnick notes, they need to explain why they can be immune to political and social influences when scientists cannot. In short, they would have to claim to be more rational than Boyle (not to mention Newton, Darwin, or Einstein).

However, such self-referential criticisms are often just shrugged off by social constructivists, and I agree with Pinnick that a more effective critique would focus upon the historical credentials of Shapin and Schaffer's analysis (Pinnick 1998: 227). *Leviathan and the Air-Pump* is an exhaustively researched work of remarkable erudition, but its focus is narrow. An obvious question is how much can be learned from one case study, even one examined in meticulous detail? Even assuming that Shapin and Schaffer got it right, is the case study typical or an aberration?

How, in general, do scientists get their methods? The Hobbes/Boyle controversy was part of a sweeping change in the way that science was done, a change fundamental to and largely definitive of modern science. Since Aristotle, the ideal of scientific knowledge had been deduction from premises that were known with certainty. Natural science thus aimed at proof and demonstration, much as the mathematical sciences do. With the rise of modern science we see the emergence of experiment as the touchstone of scientific inquiry. The results of experimental inquiry are not decisively demonstrated like a mathematical theorem, but are shown to be probable, approximate, and tentative, i.e., open to refutation by further experiment.

Most methods are far more mundane in their provenance and limited in their significance. Consider the astronomical tool discovered by Henrietta Swan Leavitt (1868–1921). How far away are the stars? In the nineteenth century, astronomers had perfected instruments of sufficient precision to measure the distances to the nearer stars by parallax, the tiny apparent shift in a star's position when viewed from opposite sides of the earth's orbit. What, though, about really distant stars, the ones too far away to exhibit parallax, even when measured by the most precise instruments? What astronomers needed was a "standard candle," an object of known intrinsic brightness. If we know how bright an object really is, we can determine its distance by how bright it looks.

A hundred years ago computers were women. Astronomical observatories needed workers who would scan and measure the images of stars on photographic plates and then do the laborious calculations of the positions and brightness of the stars. To do the work required mental stamina, oceans of patience, an obsession for accuracy, and a willingness to perform heroic labors for modest pay. Almost always, the ones able and willing to do such work were women, who were called "computers." Leavitt was one such "computer" working for the Harvard College Observatory. She found the standard candle.

Many stars vary in brightness. When the variations regularly repeat, the interval between the maxima and minima of brightness is called the "period." The first star observed to display such periodicity was delta Cephei, the fourth brightest star in the constellation Cepheus, and regularly varying stars of that sort are therefore called "Cepheids." Leavitt was a peerless finder of variable stars. In 1912 she published a remarkable result of her investigations of variable stars in the Small Magellanic Cloud. The Magellanic Clouds are relatively small companion galaxies to the Milky Way visible from the southern hemisphere (called "Magellanic" because they were first noted by Europeans during Magellan's circumnavigation).

Because the Magellanic Clouds are compact and distant groupings, it may be assumed that their stars are only a small distance from each other compared to their distances from an earthly astronomer. Therefore, those stars are all about the same distance from us. It follows that the apparent differences in the brightness of those stars are due to their intrinsic brightness, not their variable distances.[1] Leavitt noticed a precise relationship between the brightness and the periodicity of the Cepheids in the Small Magellanic Cloud. The brighter the star, the longer the period between its brightness maxima and minima. This means that the actual, not merely the apparent, brightness of a Cepheid could be mathematically correlated with its period. Therefore, if two Cepheids are observed to have the same period, but one appears a quarter as bright as the other, then we may infer that, since both have the same intrinsic brightness, the dimmer one is twice as far away (apparent brightness diminishes with respect to the square of the distance).

To turn such relative measures into absolute ones, calibration was needed, and this was eventually provided. However, even the relative distance measurements made possible by Leavitt's luminosity-period relationship were useful. Leading American astronomer Harlow Shapley used Leavitt's Cepheid scale to determine the distribution of globular clusters, dense spheres of ancient stars, and these measurements gave a strong clue to the shape of our galaxy (Asimov 1966: 59). Perhaps the most famous use of "Leavitt's Law" was when

Edwin Hubble realized that he had misidentified what he thought was a nova, an exploding star, in what was then called the Andromeda Nebula. He recognized that the "nova" was actually a Cepheid, and therefore he was able to measure its distance. It turned out to be spectacularly far away, and therefore the Andromeda Nebula had to be promoted to the Andromeda Galaxy, a distinct "island universe" far separated from the Milky Way.

How well does Henrietta Leavitt's discovery of the period-luminosity curve fit with the Shapin and Schaffer account of how scientific methods originate? Not very. Leavitt had no political advantages. She was a woman at a time when the sciences were overwhelmingly dominated by men. Men occupied all of the prestigious positions, leaving women to take the lower-status "computer" jobs. When Leavitt's discovery was published in 1912, her boss, Charles Edward Pickering, insisted that he be listed as author, though it was general knowledge that it was Leavitt's work (Clark and Clark 2004: 97). In later years, Shapley paid Leavitt the backhanded compliment of seeming to hint that most of the credit for her discovery belonged to him (Johnson 2005: 119–120). When a woman makes a discovery that is important enough that two higher-status men want to claim credit for it, she must be on to something. Further, there were no ideological or religious issues at stake in the adoption of the method. There seems to be no other reasonable explanation for its use other than that it is, in fact, a reliable way to gauge stellar distances.

What, after all, could be socially constructed in this case? The brightness of stars and the periods between their maxima and minima could be accurately measured and correlated by anyone, like Leavitt, with a keen eye, a sharp mind, and plenty of determination. Many scientific methods are like this, deriving from the discovery of simple empirical correlations. Others are provided by improvements in instruments, or the development of more sophisticated mathematical tools. Any scientific field such as astronomy will employ many different methods at any given time. These often come and go, and scientists debate methodology just as vigorously as anything else. Such debates are part of the ordinary, routine process of science, and it is implausible in the extreme to think that each turns on heavy-duty political, ideological, and religious issues.

Perhaps, though, Shapin and Schaffer would reply that the methodological changes that interest them are the *really big* ones, the ones that redefined the nature of science. It is these that must have political, ideological, and sociological explanations, as their history attempts to show.

In general, then, how does *Leviathan and the Air-Pump* rate as a work of history? Pinnick thinks that Shapin and Schaffer create a false dichotomy by overstating the opposition between Hobbes and Boyle (Pinnick 1998: 231–236). In fact, she claims, as their writings on methodology show, Hobbes was friendlier to experiment and Boyle more accepting of deductive demonstration than Shapin and Schaffer indicate; there was no mutual exclusion. Therefore, the sharp methodological divisions that are required to justify Shapin and Schaffer's conclusion just did not exist. Historian John H. Zammito regards Pinnick's claim as "*Balderdash*" (Zammito 2004: 170; emphasis in original). Zammito says that at that time there really was a shift in scientific methodology from emphasis on certainty and deductive logic to experimentalism and the embrace of fallibilism and probability.

Pinnick does not deny that such an epochal shift in methodology occurred in the seventeenth century. She questions that the Hobbes/Boyle debate was as central to this controversy as

Shapin and Schaffer presume (Pinnick 1998: 231). Zammito endorses Shapin and Schaffer's focus on the Hobbes/Boyle dispute:

> The emergence of a fallibilist, probabilist approach to knowledge of the natural world is an epochal shift, and locating it in the context of the Hobbes-Boyle controversy is historically illuminating because it crystallizes the issues of a longer term problem into a self-conscious moment of controversy. Boyle and his associates self-consciously abandoned absolute deductive argumentation in natural science . . . and opted instead for an avowedly contingent and fallible approximation.
>
> (Zammito 2004: 173–174)

However, to say that the Hobbes/Boyle dispute was "historically illuminating" because it "crystallizes" a large-scale shift does not refute Pinnick's claim that Shapin and Schaffer assume a more central and decisive role for the controversy than is justified.

Interestingly, Pinnick, the philosopher, thinks that the problems with *Leviathan and the Air-Pump* are mainly historical, and Zammito, the historian, thinks that they are primarily philosophical. Zammito asserts, correctly in my view, that the main philosophical problem is that Shapin and Schaffer rely reflexively and uncritically on the underdetermination thesis (Zammito 2004: 180). They take it for granted that all experimental results are underdetermined and that there therefore can be no compelling reasons to accept any particular result. Yet we saw previously that some experimental results are so robust that it can be perverse not to accept them. When those unwilling to accept certain plain and repeatable outcomes tie themselves into conceptual knots, and resort to rhetoric and increasingly desperate ad hoc excuses, their colleagues rightly judge that they are being willfully obtuse. So, Shapin and Schaffer view their historical materials through the lens of philosophical bias.

As for historical bias, Gross and Levitt have a point when they note that Shapin and Schaffer overlook an obviously relevant historical detail, namely that there was a perfectly sound apolitical and non-ideological reason that Boyle and his allies dismissed Hobbes (Gross and Levitt 1994: 66–67). Hobbes was a crank. He carried on fatuous and shameful disputes with qualified mathematicians in Boyle's circle, long after his mathematical incompetence had been plainly demonstrated. Zammito objects that Gross and Levitt, whom he notes are not professional historians, assert that Hobbes' mathematical howlers were "indisputably the decisive factor" in the rejection of Hobbes (Zammito 2004: 171).

However, Gross and Levitt do not say this. They say that the relevance for historians of Hobbes' mathematical foolishness is a "central" question and that his standing as a crackpot was a "concrete and substantive" reason for the experimentalists' disdain (Gross and Levitt 1994: 66–67). That Hobbes' failures as a geometer "can hardly have been irrelevant" (Gross and Levitt 1994: 67) seems obvious, and even those not professional historians should be allowed to point out the obvious. Finally, Gross and Levitt plausibly attribute Shapin and Schaffer's inattention to this point to ideological blinders, the dogged determination to explain exclusively in terms of political and social imperatives (Gross and Levitt 1994: 68).

My assessment is that Pinnick and Zammito are both right: *Leviathan and the Air-Pump* is philosophically dogmatic and historically simplistic. As Zammito (Zammito 2004: 180) said, the authors assume that underdetermination shows that all experimental outcomes are equally credible, or at least that none is rationally conclusive. As Pinnick noted, they also assume that the Hobbes/Boyle debate was pivotal in the epochal change from a deductive to an experimental methodology. Yet large-scale changes in human thought are complexly caused, and no one particular incident or controversy can be singled out as *the* cause. Epochal changes usually take years or even decades to be thoroughly worked out, and the working-out is a convoluted process. The Reformation was far more than Luther's defiance at the Diet of Worms, and the rise of experimental science was far more than the Hobbes/Boyle debate. Yet Shapin and Schaffer's whole case rests on the alleged centrality of that debate.

An adequate history of the rise of modern experimental science would have to cite an intricate multiplicity of diverse factors, some rational and evidential, and, no doubt, some political and social (See Wootton 2015: 310–360, for such a balanced account that shows the complexity of the events that—eventually—shaped the consensus on experimental methods.). Shapin and Schaffer are committed to a particular mode of explanation, having assured themselves *a priori* that the history of science must be explained *only* in terms of politics and social alliances. This commitment skews their judgment and makes them ignore obviously relevant points, as Gross and Levitt note. Therefore, *Leviathan and the Air-Pump* is deficient as an account of the Hobbes/Boyle dispute and, *a fortiori*, as a general account of how scientific methods arise.

As with relativism, much more could be said about social constructivism. However, after examining the futility and failure of some of its chief works and practitioners, the project must appear dubious.

So, in this chapter and the previous one, we have looked at several serious challenges to scientific rationality that were made during the science wars. Powerful replies to those challenges were made at the time, and I have added my support to those rejoinders. Of course, in two chapters we can only scratch the surface of very complex debates, and the bibliography will offer guidance to in-depth discussions. However, I think that enough has been said to indicate that the high reputation and the trustworthiness of science, and the emptiness of some of the main criticisms leveled against it, emerge as clear conclusions from the science wars.

Thomas Kuhn: Foe of Science?

3

The Ashtray Incident

Under the best of circumstances, the graduate student's life is not a happy one. Overworked, grossly underpaid, and powerless, the doctoral student must endure several years of what one of my professors at the University of Pittsburgh called academic "boot camp." Your life is made much worse if you are under the thumb of a bad supervisor. According to Errol Morris, Thomas Kuhn was one of the worst. Morris, who later gained fame as a movie director, was in the early seventies a graduate student in Princeton's Program in the History and Philosophy of Science. Kuhn was the head of that program. According to Morris, Kuhn was imperious, intemperate, dogmatic, and, on one occasion, physically violent (see Morris 2018: 7–14).

The incident occurred when Morris was in Kuhn's office discussing Kuhn's angry and abusive response to a paper submitted by Morris. Morris had written on James Clerk Maxwell, the Scottish physicist of the nineteenth century famed for his work on electromagnetism. Apparently, Kuhn considered Morris's work to be "whiggish." The cardinal sin for a historian is to write Whig history, that is, history that fails to take past agents on their own terms, but judges them by anachronous standards derived from present beliefs and values. Whig history of science, for instance, tends to divide past scientists into good guys and bad guys where the good guys were the ones seen as forerunners of present science while the bad guys are those who opposed the trends leading to the present. For instance, a Whig historian would praise Charles Darwin as a great discoverer, but castigate Darwin's antagonist Richard Owen as an obscurantist.

As a recommendation of historical sensitivity, the practice of interpreting historical events in context, the strictures against Whig history are unexceptionable. However, Morris sensibly asked whether, due to his doctrine of "incommensurability" (see below), Kuhn was not in fact required to hold that past theories are incomprehensible on their own terms, meaning that if we discuss them at all it can only be on our terms. Thus, did not Kuhn's own theory entail that our interpretations of past theories must be whiggish? This question apparently so angered Kuhn that he flung a heavy ashtray at Morris.

The missile missed, but Kuhn threw Morris out of Princeton. In 2018, over 45 years after the ashtray incident, Morris, clearly still angry, published *The Ashtray (Or the Man who Denied Reality)*, which can only be described as a philosophical exposé. In this entertaining, informative, and lavishly illustrated book, Morris not only depicts Kuhn himself in a bad light, but judges his most famous book, *The Structure of Scientific Revolutions*, to be murky in its ideas, unoriginal in content, and pernicious in its aims—a tedious argument of insidious intent. As his subtitle implies, Morris says that Kuhn denied the existence of objective external reality, and so also repudiated the idea that our assertions are made true by correspondence with that objective reality. To say that the cat is on the mat is not made true by the cat actually being on the mat; indeed, if there is no determinate external reality, then there is no cat and no mat apart from our representations of them. Morris regards such ideas as bizarrely wrong and dangerous. Since *Structure* first appeared in 1962 some critics have regarded it with frank loathing and have seen Kuhn as a proponent of irrationality.

Contrast such judgments with that of eminent philosopher of science Ian Hacking who begins his introductory essay to the fiftieth-anniversary edition of *Structure* as follows: "Great books are rare. This is one. Read it and you will see" (Hacking 2012: vii). Other encomia could be quoted, but the best measure of a book's influence is the degree to which later writers defer to it. As we will see in the fourth chapter, when challenged about her view of science as incapable of achieving objectivity as normally construed, Sandra Harding invoked the authority of Kuhn. For all of the social constructivists, sociologists of knowledge, postmodernists, and other radical science critics mentioned in the previous chapters, all roads lead back to Kuhn. Indeed, Kuhn's influence, due to *Structure*, has extended to many academic fields. As detailed by philosopher James A. Marcum, Kuhn's ideas have had enormous influence on the behavioral, social, and political sciences, as well as science policy, science education, and legal studies (Marcum 2015: 201–231).

WHY?

So, for some, Kuhn's book is pernicious claptrap. Others regarded the book as an epiphany, a work that liberates us from stultifying, oppressive, and blinkered assumptions about science and human rationality. Kuhn himself seemed taken aback by the reactions, and spent much of the rest of his life fending off both what he perceived as the unfair characterizations of critics, and the equally distorting misconstructions of would-be admirers. Why the fuss? The book is hardly a potboiler. It is a serious and rather technical study of the historiography and history of science. Why such a range of intense reactions? The short answer is that Kuhn appeared to release the genie of relativism.

Friends and enemies took Kuhn to be saying that—when examined in detail and without distorting whiggish assumptions—revolutionary episodes in science cannot be understood in terms of the same sorts of rational norms that guide us during periods of "normal science." During such "normal" periods, science is guided by a ruling "paradigm," that is, a preeminent scientific achievement that specifies both the sorts of problems researchers in that field are expected to address and the general form of the answers that are anticipated. Paradigms, as exemplars of the way good science is to be done, imply certain norms, standards, values, and methods that are definitive of good scientific practice at that time.

Yet revolutionary changes in science are of a very different sort than the routine "problem solving" that occurs when a discipline is firmly under a reigning paradigm. In revolutionary episodes, the paradigms themselves are rejected and replaced by new paradigms. Such paradigm shifts apparently involve wholesale change not only in theory, but in the very norms, values, and evidence whereby theories are assessed. Kuhn likened the change to a "gestalt switch," whereby there is a sudden holistic change of perspective, as the famous duck/rabbit illustration can switch between appearing as a duck and appearing as a rabbit.

How, then do we say that paradigm shifts are norm-guided, rational enterprises? Are they not instead "world changes" whereby an entire constellation of "truths," meanings, values, standards, methods, and observations is dropped in favor of an entirely new constellation? The change is total, so sweeping and comprehensive that there is no background set of concepts and values that permit us to judge that the shift was a rational and progressive procedure. One critic said that for Kuhn paradigm change is due to "mob psychology" (Lakatos and Musgrave 1970: 178). Looked at broadly, therefore, is not the lesson of the history of science that, over time, we do not know more, we only "know" differently?

In fact, adherents of different paradigms will find that their discourse is now "incommensurable" with each other. They inevitably, at least to some extent, talk past one another since their concepts do not directly translate into the idiom of the other paradigm. One is talking about apples and the other is talking about oranges even if they use the same words to pick out their concepts. "Mass" meant something very different for Newton than for Einstein.

So, Kuhn's analysis seemed to many to imply that the very history of science apparently justifies conceptual relativism. Further, theory choice appears irrational since it is not guided by neutral norms or objective evidence. Successive ages conceptualize the world differently, but no neutral criteria exist that permit us to say that one age's understanding is closer to reality than another's. All such standards are paradigm-bound and so cannot, without begging the question, be employed to judge between paradigms. "True" becomes a predicate that can be deployed only within paradigms, not between them. The idea of achieving convergence between our theories and reality becomes empty since our "reality" will be defined by our paradigms. Such thoughts were music to the ears of alienated literary intellectuals but strident cacophony to scientists and many philosophers of science.

Was Kuhn a relativist? Did he imply that major changes in theory are irrational? Ian Hacking calls such charges "absurd" given Kuhn's later explicit denials (Hacking 2012: xxxi).

I think, however, that we have to be more charitable to those, who, either scornfully or approvingly, read the Kuhn of *Structure* as a defender of relativism or as making a case for the irrationality of theory choice. For one thing, many very smart people read him that way, which makes it hard to think of such an interpretation as just foolish. Hacking is right that in later writings Kuhn does disavow relativism and affirms certain rational standards of theory choice. However, actions speak louder than words, or, in the case of written products, some words speak louder than others. Those who say provocative things when young might have trouble trying to un-ring the bell when older.

One of the problems Kuhn presents would-be interpreters is that he writes in a semi-oracular style. Some have said that *Structure* is written in aphorisms. Such a style can be stimulating, but it has drawbacks for readers who seek precision and clarity. Indeed, I would conjecture that it is the obscurity of Kuhn's style that accounts for much of the popularity of his book. The human mind abhors a vacuum of meaning, and when confronted with vagueness or ambiguity, people spontaneously and unconsciously project their own meanings to fill the void. Interpreters then regard their own projections as the meaning of the text, which, of course, they then judge to be insightful.

Another problem is that Kuhn often makes hair-raising pronouncements but follows them with apparently moderating language. It is hard to know which to take more seriously, the part where he says it or the part where he walks it back. As physicists Alan Sokal and Jean Bricmont note, there seem to be two Kuhns—the moderate Kuhn and the immoderate Kuhn—who appear to make very different claims about rationality (Sokal and Bricmont 1998: 75). Similarly, philosopher

A.F. Chalmers identifies two incompatible strands in *Structure* and its Postscript, a relativist strand and a non-relativist one (Chalmers 2013: 115).

Because Kuhn presents these two aspects or faces, I will divide my treatment of him into two parts. In the remainder of this chapter, I will consider Kuhn interpreted as a radical/relativist/irrationalist. I will ask whether the Kuhn of *Structure* effectively undermines the view of scientific revolutions that would be held by a scientific rationalist as defined in the previous chapter. In the next chapter, we will consider the moderate/non-relativist Kuhn, and consider some of his post-*Structure* writings and how they modified or qualified his view.

For the scientific rationalist, revolutionary changes in science are *warranted*, that is, unbiased evidence and sound arguments were available that were sufficient to make the scientific community's shift to the new view eminently rational. No rhetorical grandstanding, circular arguments, or propaganda were needed. The impartial reasons were *there*, and key observations and evidence really did authorize the claims of the new theory over the old. Further, in these revolutionary episodes, the evidence really mattered, that is, it played a crucial role in the historical shift to the new view. The rationalist need not deny that psychological and sociological factors can influence the course of science. Clearly, they do. However, the rationalist will insist that *reasons* are what mattered most to convince the members of scientific communities to adopt new paradigms.

To see whether and to what extent Kuhn contradicts the above view of scientific revolutions, I now turn to a consideration of some main themes of the original 13 sections of *Structure* published in 1962. Kuhn lived until 1996 and was quite productive during those latter 34 years. Why, then, focus on these sections and not later writings also? Because the 1962 publication was the bombshell. It was the manifesto that got things stirred up. Later writings clarify, modify, qualify, and, perhaps, retreat from things asserted in 1962. Kuhn's later thoughts may have been better—or not—but it was his first thoughts in sections I through XIII of Structure that made it one of the most influential books of the century, and which, I submit, account for the preponderance of Kuhn's continuing influence. So, these sections are my focus here.

At this point it would be appropriate to have a detailed summary of sections I-XIII of Structure. However, since many readers may already be quite familiar with these texts, the summary will be placed in the online companion to this book. Those who have not read Kuhn's book, or those who would like a refresher, are *strongly* encouraged to read the summary before proceeding. Page numbers cited below will refer to the fiftieth-anniversary edition of Structure, which are the same for the Third Edition.

KUHN VS. RATIONALITY?

Inevitably, and quite understandably, Kuhn's arguments sounded to many readers like a radical form of relativism or irrationalism. Scientific methods, standards, and values are determined by paradigms, and not neutral norms for judging between them. New theories are not accepted by being compared *vis-à-vis* impartial evidence, but by a sort of gestalt-switch-like experience whereby the new theory is accepted via a radical and comprehensive change of perspective, a "conversion." Worse, once such a fundamental change of vision had occurred, the converts no longer even lived in the same reality as before but now should be spoken of as residing in a different world. Science may progress in a sense, but it is not a progress towards truth. If this is not a defense of relativism or irrationalism, it is an excellent impersonation of one.

Yet distinguishing between what Kuhn means and what he only seems to mean is not easy. As noted, Kuhn often pairs bold statements with seeming qualifications. Also, it is hard to know what to take literally and what to regard as a figure of speech. I would like to regard talk about "world changes," "incommensurability," and "gestalt switches" as elaborate metaphors. There is no question that after a major conceptual revolution things look very different. Such comprehensive changes in our understanding may be intimidating, as the new science was for John Dunne:

The element of fire is quite put out;
The Sunne is lost and th'earth, and no
Mans wit Can well direct him
Where to looke for it.

Or the changes may be exhilarating, as Alexander Pope expressed in his famous couplet:

Nature and Nature's Laws lay hid in Night.
God said Let Newton Be! and all was light.

Perhaps Kuhn is only employing vivid metaphors to evoke that sense of living in a new world, one that is thrilling or disturbing, after a comprehensive change in our thinking.

I would like to read Kuhn this way, but that would fail to take him seriously. Whatever his qualifications, rhetorical flourishes, and walk-backs, he plainly means it when he says, for instance, that Galileo's visual experience was different from Aristotle's, and that the "world of the professional astronomer" (whatever that means) had fewer stars and one more planet after the discovery of Uranus, and that after the work of Benjamin Franklin, a Leyden jar *became* a condenser. To say that by these expressions Kuhn only meant to say that scientists' beliefs had changed, and did not mean that Galileo literally saw something different, or that the universe really had fewer stars, or that the device was literally changed into a condenser by scientists' perceptions—would be to dismiss Kuhn and render his claims trivial. Everybody says that scientists' beliefs changed on those occasions. If that were all that Kuhn meant to say, then Structure should have been greeted with yawns rather than excitement and controversy.

What, then, are we to make of Kuhn's claim that the *visual* experience of Galileo differed from Aristotle's (121) Kuhn denies that Aristotle and Galileo merely interpreted what they saw differently (122). Kuhn is aware of the difficulties involved with saying that when Aristotle and Galileo looked at swinging stones, the former *saw* constrained fall and the second *saw* a pendulum (122). Yet he insists that we must make sense of statements like these (122). They are not merely metaphorical. They have a literal meaning, but what?

Does Kuhn mean that Aristotle saw the swinging stones *as* an instance of constrained fall while Galileo saw it *as* a pendulum? However, to say "P saw a y" cannot mean the same thing as "P saw an x as a y." To say that P saw a y implies that what P saw actually was a y, whereas "P saw an x as a y" carries no such implication. Kuhn is aware of this difference between seeing and seeing-as and denies that he is endorsing the "seeing as" account: "Scientists do not see something *as* something else; they simply see it" (85). Galileo saw a pendulum; astronomers after Herschel saw a new planet; Lavoisier saw oxygen; Franklin saw a condenser. Their world had changed.

Such language raises many questions. Does Kuhn mean that Galileo saw a pendulum in the same sense that he saw the moons of Jupiter? If there was no pendulum until he saw it, would Kuhn say that, by that same measure, Jupiter had no moons until Galileo saw them? At the end of the next chapter we will see that the answer to this question may be "yes."

Instead of devoting many pages to the daunting effort to untangle just what Kuhn might have meant, I will cut to the chase: Does Kuhn *clearly* say anything in *Structure* that would preclude the deployment of objective, non-question-begging observational evidence in the debates over revolutionary changes in science? *Prima facie*, at least, he does. Kuhn says that proponents of different paradigms *necessarily* talk through each other and that experiment is just irrelevant in settling their differences (132). Aristotle and Galileo were just fundamentally at cross-purposes (132), and in such a situation, as in a political revolution, the two sides can only deploy circular arguments and techniques of mass persuasion against one another (94).

Is this really so? There is no question that scientists sometimes talk past each other and sometimes resort to rhetoric, propaganda, or question-begging rather than logical argument. But do they *necessarily* do so? Would Galileo have *nothing* to say that should count as a good argument or a relevant observation for Aristotle?[1] Of course, Galileo spent much of his career arguing with the Aristotelians of his day. Were all of his arguments circular and all of his observations irrelevant? Let's consider some of the arguments made in Galileo's *Siderius Nuncius*, the *Starry Messenger*.

In 1609, Galileo heard a report of a new invention, a spyglass that could make things far away look closer. Soon he had constructed his own telescope of twenty power, the best in the world at the time (though far inferior to telescopes available to any amateur observer today). In November and December of 1609, he observed the Moon as it went through its phases. What he saw directly contradicted the prevailing Aristotelian account. According to that view, the Moon, as a heavenly body, was supposed to be perfectly smooth and spherical, consistent with the perfection of the heavens as opposed to the chaotic topography of the corrupted terrestrial sphere. Historian Albert van Helden describes some of these observations:

> When Galileo examined the Moon with his twenty-powered spyglass, its surface appeared anything but smooth: it seemed rough and uneven. The dividing line between light and darkness (the terminator) was not a smoothly curved line at all, as one would expect if the Moon's surface were perfectly smooth. Instead it was very irregular. In the bright part features were starkly outlined by black lines that grew broader and narrower as the light of the Sun varied; in the dark part there were little patches of light. Galileo drew the conclusion that the Moon's surface is full of mountains, valleys, and plains, just as the Earth's surface is.
>
> (Van Helden in Galileo 1989: 11)

These and many other observations were presented in Galileo's brilliant little book. Perhaps Galileo's most famous discovery was the four "Galilean" moons of Jupiter—Ganymede, Europa, Callisto, and Io. *Prima facie*, these observations were solid, non-circular evidence against the Aristotelian view—evidence, not rhetoric, propaganda, or mere persuasion.

Of course, conservative Aristotelian astronomers objected, and their most potent objection was that Galileo did beg the question by making insupportable claims for his telescope. Galileo's Aristotelian opponents placed a special emphasis on the reliability of the unaided senses (Chalmers 1990: 52–53). The Aristotelian/Christian tradition held that the senses have the function of informing us about the world, and when carefully used in the appropriate circumstances, and if not impaired by illness or inebriation, the senses are our trustworthy guides. Galileo, however, said that his spyglass was more reliable than unaided vision for astronomical observation, and this contradicted a deeply ingrained epistemological standard. As Kuhn says, new paradigms bring new standards. So, was Galileo not begging the question by appealing to his standards, and not those of his opponents?

A.F. Chalmers notes, however, that Galileo offered strong arguments for the veracity of his observations of Jupiter's moons, although those moons are invisible to the naked eye:

> He could, and did, argue against the suggestion that they [The Jovian moons] were an illusion produced by the telescope by pointing out that the suggestion made it difficult to explain why the satellites appeared near Jupiter and nowhere else. Galileo could also appeal to the consistency and repeatability of his measurements and their compatibility with the assumption that the satellites orbit Jupiter with a constant period. Galileo's quantitative data were verified by independent observers, including observers at the Collegio Romano and the Court of the Pope in Rome. What is more, Galileo was able to predict further positions of the satellites and the occurrence of transits and eclipses, and these too were confirmed by himself and independent observers.
>
> (Chalmers 1990: 55)

Astrophysicist and popular science writer John Gribbin notes that a committee of Jesuit scholars appointed by Galileo's later nemesis Cardinal Bellarmine confirmed the following of Galileo's observations:

i. The Milky Way really is made up of a vast number of stars.
ii. Saturn has a strange oval shape, with lumps on either side [Better telescopes showed the "lumps," of course, to be rings—KMP]
iii. the moon's surface is irregular
iv. Venus exhibits phases
v. Jupiter has four satellites (Gribbin 2002: 90).

So, yes, Galileo did propose new a new standard for astronomical observation. A better one. His claims for the new standard were confirmed by the Pope's own astronomers.

Kuhn therefore exaggerates the difficulty of presenting objective observational evidence for new paradigms. Further, old paradigms do not so indoctrinate and condition their adherents that they are rendered incapable of seeing the objective evidence for the new one. Observation may indeed be "theory-laden" in the sense that our sensory experience is not passive reception of raw data but active interpretation in the light of all we have been taught, including the theories we accept. A literate person looking at a text sees words, not merely shapes arranged against a background, which is presumably what a wholly non-literate person would see. However, the impact of our theories on our perception is not so strong as to preclude seeing the unexpected. This point is decisively stated by philosopher Tim Maudlin:

> If presented with a moon rock, Aristotle would experience it as a rock, and as an object with a tendency to fall.

He could not fail to conclude that the material of which the moon is made is not fundamentally different from terrestrial material with respect to its natural motion. Similarly, ever better telescopes revealed more clearly the phases of Venus, irrespective of one's cosmology and even Ptolemy would have remarked the apparent rotation of the Foucault pendulum. The sense in which one's paradigm may influence one's experience of the world cannot be so strong as to guarantee that one's experience will always accord with one's theories, else the need to revise theories would never arise.

(quoted in Sokal and Bricmont 1998: 75–76)

What, then, about incommensurability, often regarded as Kuhn's most serious challenge to rationality? "Incommensurability" takes various forms for Kuhn. There is incommensurability of values, of standards, and of meanings. However, it was the claim of incommensurability in the sense of radical meaning variance—that identical terms in competing paradigms change in meaning and reference—that seemed to provoke the most controversy. Therefore, it is this sense of "incommensurability" that I will focus on here.

I would like to begin by saying what incommensurability is *not*. First, incommensurability cannot be a breakdown of communication due to the fact that the opposing parties are too angry or alienated to listen to each other. It was not that North and South *could not* communicate just prior to the Civil War; they *would not*. Neither can incommensurability be a failure of communication due to a simple misunderstanding of terms. Talking past one another can be quite frustrating, but it can be remedied by specifying what we are talking about. If, on the other hand, terms cannot be clarified because they are meaningless or incorrigibly vague or ambiguous, this is not incommensurability either. Finally, incommensurability cannot be invincibility to argument or evidence due to an absolute commitment to an ideology or creed. When no evidence can convince an ideologue, this need not be due to a breakdown of communication, but to the impenetrability of an ideological shell. The ideologue is psychologically incapable of fairly considering the arguments and evidence for the other side.

To be philosophically interesting, incommensurability has to be a *semantic* matter. It has to be an *irremediable* and *in-principle* breakdown in communication due to an irresolvable linguistic incompatibility. If my discourse is incommensurable with yours, it must be impossible in principle for us to express what we mean so that the other will find it fully comprehensible. Thus, despite the best will to communicate and their best efforts to clarify terms, proponents of different views find that their ability to achieve mutual understanding has irreparably broken down, at least in part. Further, and crucially, to have import for the rationality of science, the breakdown in communication between proponents of two rival theories must be such as to impair the rational comparison of the two competing theories.

What, then, is the alleged problem of incommensurability? It is hard to say since how Kuhn characterizes the concept changes rather maddeningly over time (Khalidi 2000: 171). What Kuhn says about the concept in section XII of *Structure* is rather sparse. He says that proponents of different paradigms unavoidably partly talk through one another and fail to make "complete contact" with each other's viewpoints (147). The reason is that the meanings of theories and their constituent terms have to be grasped as a whole, and cannot be rightly understood if extracted from that context and placed in the foreign environment of a different theory. Thus, only those who have experienced the "conversion" to a new paradigm will be able to understand completely the meaning of the new theory, and so communication across a revolutionary divide is "inevitably partial" (148). Not only do the meanings of theoretical terms change across paradigms, but even the meaning of observables such as "earth" and "motion" (149). So fundamental are these changes of meaning that proponents of different paradigms are said, in some literal but difficult to specify sense, to live in different "worlds" (149).

Kuhn later complained that philosophers (and only philosophers, he said) had misunderstood these remarks (197). Again, however, if there was misunderstanding, it was excusable and probably inevitable. It is hard to read these passages without getting the impression that for Kuhn those who attempt to communicate across paradigms will be frustrated because they will experience some degree of unavoidable incomprehension, misunderstanding, or misinterpretation. If this is a misreading, then it has been a very influential misreading, and needs to be addressed on its own terms even if Kuhn later repudiated it.

Taking the claims in these passages as straightforwardly as I can, they seem to me not to deserve the volumes of philosophical commentary they have drawn. As Morris implied in the comment that provoked the flung ashtray, Kuhn does not appear to see himself as suffering from incommensurability. He describes in great detail the fine points of past theories and exactly what their terms meant and their intended referents. As prominent philosopher Hilary Putnam notes in his interview with Morris, to say that past theories are incommensurable with present ones, and then to claim to be able, from our perspective informed by present theories, to explicate the details and nuances of those past theories—is just incoherent (Morris 2018: 52). If Kuhn can understand and articulate the meanings and intended references of terms in "incommensurable" theories—without whiggish distortion—then so can scientists.

Further, as Putnam also points out in his interview with Morris, to regard even the terms applied to the most common observables (e.g., earth, motion) as totally changing meaning when our theories change is patently absurd. It entails, for instance, that since we now know that grass lives by photosynthesis, we cannot understand what "grass" means in eighteenth-century literature (Morris 2018: 53). Wordsworth's line about "splendour in the grass" should make us scratch our heads and puzzle "splendour in the what?"

In general, do proponents of very different theories find that their ability to communicate has unavoidably broken down, at least in part? Do very different ideas attached to key terms to any degree impair mutual understanding or rational debate? Is this always or typically what happens? Consider the disagreements between Charles Darwin and his bitterest critic Richard Owen and their very different understanding of the key shared term "homology."

Owen, the leading anatomist of his day, noted that the skeletons of very different animals evince a deep similarity, and that the bones of one species could often be matched one to one with the bones of another. Owen called these corresponding traits "homologies." Darwin seized this fact, and appropriated the term, using it as a main item of evidence for evolution. If we see that the hand of a human, the wing of a bat, the flipper of a dolphin, the paw of a bear, and the leg of a horse have corresponding bones, though modified to function very differently, what should we infer? For Darwin it was plain: The corresponding structures were inherited from a common ancestor of the deep past, but modified by natural selection to serve very different purposes. The human hand is used for grasping, the dolphin's flipper for swimming, and the bat's wing for flight.

Owen, the staunch antievolutionist, held that the similarities identified as homologous were reflective of "archetypes," ideal patterns of organic forms, something like the Ideal Forms of Plato, that served as the generalized blueprint of organisms (Desmond 1982: 42–44). What Darwin understood in terms of descent with modification, Owen understood in terms of ideal essences. Obviously, then, Owen and Darwin took very different theoretical stands. Did this mean that they understood the meaning of "homology" in mutually incomprehensible ways?

Not necessarily, as Michael J. Donoghue explains:

> Owen's definition [of "homology"] was more or less retained by Charles Darwin . . . and his contemporaries—homology was morphological correspondence as determined primarily by relative position and connection.
> (Donoghue 1992: 170)

So, Darwin, at first anyway, understood "homology" in much the same way as Owen, as a correspondence of anatomical similarities. However, Donoghue continues:

> But the associated explanation was radically different. Metaphysical archetypes and "essential" similarity were

> replaced by material ancestors that could evolve. The connection having been made, the observation of homology could then be used as evidence for evolution . . .
>
> (Donoghue 1992:170–171)

Perhaps, then, where Kuhn sees incommensurability of meaning, this might on many occasions be more plausibly understood—in an old-fashioned, pre-Kuhnian way—as agreement in meaning but a difference in the *theoretical understanding* or *explanation* of the referents picked out by that meaning (e.g., with respect to "homology," the interspecific anatomical similarities).

If, as Kuhn claims, there *are* cases where the meaning and reference of observational terms really do change radically between competing theories, then the consequence would be that proponents of such competing theories cannot offer different theoretical understandings or explanations of the same thing. When they appear to do so, they would only be equivocating because they are attempting to understand or explain different things, even if they are using the same terms. In this case, however, the "opposing" theories would not really be rivals since each would have a different subject matter. Yet then the whole notion that the theories could be incommensurable breaks down. As John Preston notes, for two theories to be incommensurable, they have to share the same subject matter, that is, they have to be *about* the same things. It would be silly to say, for instance, that quantum theory and evolutionary theory are incommensurable; they are about totally different things (Preston 2008: 92). However, if it is possible to specify a common domain for two allegedly incommensurable theories, then there must exist a language to express these commonalities of reference and meaning. That is, there must exist a neutral language that can express each theory, revealing their agreements and disagreements about a shared subject matter (Preston 2008: 92). In that case, the claim of incommensurability will be self-refuting.

In general, judging by the copious historical evidence, were Darwin and Owen in any sense condemned to mutual misunderstanding? Was there any sort of unavoidable and irremediable breakdown in their ability to communicate or to offer objective evidence and rational arguments to each other? No, not even partially. Darwin did not reply to Owen's critique, leaving the fireworks to friends like the terrifying polemicist T.H. Huxley. In his letters to colleagues Darwin complains that Owen had maliciously misrepresented him and had criticized him unfairly. However, there is no evidence whatsoever that Owen *could not* have understood Darwin. On the contrary, the whole basis of Darwin's complaint is that Owen *could* have understood him but spitefully *would* not. Further, though Darwin clearly resented Owen's criticisms, there is no reason to think that he failed to comprehend them. Rather, he got the points all too well (on Owen's criticism of Darwin and Darwin's reaction see Hull 1973: 171–215).[2]

Suppose, though, that Donoghue is wrong and that, as Ernst Mayr claims, since the publication of *The Origin of Species*, in 1859, there is only *one* definition of "homology" that makes biological sense (quoting George Gaylord Simpson, with interpolations):

> A feature [character, structure, and so on] is homologous in two or more taxa if it can be traced back to [derived from] the same [a corresponding] feature in the presumptive common ancestor of these taxa.
> (Mayr 1982: 232; Simpson 1975: 17)

The antievolutionist Owen of course could not define "homology" in terms of derivation from a common ancestor. Would this mean that since 1859 an evolutionist and Owen would have to just talk past each other? No, and the reason is that the crucial question is not whether the terms of *theories* change in meaning but whether there is a breakdown in understanding between *theorists*. Contrary to what Kuhn seems to think, the former does not entail the latter, not even partially. Being a Darwinian cannot preclude you from understanding what Owen meant by defining "homology" in terms of ideal archetypes. Darwinians will thank Owen for recognizing the fundamental morphological correspondences, but will regard the conceptualization of them as reflecting ideal archetypes, in Mayr's terms, as making no biological sense.

Scientists can, and frequently do, change the meanings of terms due to theoretical advances, and those changes are then argued out. A famous example is the change in the definition of "planet" that led to the demotion of Pluto to "dwarf planet" status. The change was prompted by discoveries that led to significant revisions in how solar system bodies are conceived, changes that indicated that Pluto is a different kind of object than Earth or Jupiter. The debate over the change was vigorous, even acrimonious, but the opposing sides *debated*, and if the disputants sometimes miscommunicated, there is no reason to think that they were forced to do so by unavoidable semantic necessity. (For an update on the debate, see Lisa Grossman, "Pluto's Place," *Science News*, Vol. 200, #4: 20–23.) Definitions are open for scientific debate as much as anything.

Speaking personally, I have two fundamental objections to Kuhn's account in *Structure*: (1) It is simplistic and overly schematic, and (2) it ignores the main driver of scientific change—the natural world itself.

First, by explaining, literally, nearly everything about science in terms of paradigms, Kuhn grossly overgeneralizes and oversimplifies. ALL human activities, including science, are *complexly* caused by a great heterogeneity of factors. To explain virtually the entire nature of a science in terms of a one-size-fits-all concept like "paradigm" is to ignore, *inter alia*, innumerable contingencies of time and place such as personal eccentricities and idiosyncrasies—and individual genius, serendipity (the invention of the telescope changed astronomy more than any purely conceptual change), unpredictable and seemingly improbable connections between ostensibly unrelated things (such as those featured in James Burke's BBC documentary *Connections*), sheer luck (e.g., chance discoveries, like penicillin), and ideology backed by terror (as when Stalin imposed Lysenkoism on Soviet geneticists). If "paradigm" is made so protean a concept as to cover all of these factors, then, as many critics have charged, it is emptied of any distinct meaning.

Finally, and most importantly, *nature* plays the biggest role in setting our scientific agendas. Though Kuhn later (sort of) conceded a role to nature in the formation of scientific knowledge (see Marcum 2015: 139–140), in *Structure* such a role, if present at all, has shrunk to a geometric point. And this is a problem. To the best of our knowledge, some things have just been *found* to be so, and, since their discovery, have been accepted as facts by every competent practitioner, and have endured *through* major theoretical upheavals in thought: The blood circulates. The earth moves. Water is H_2O. Evolution occurred. Germs cause disease. DNA has a double helix structure. Nuclear fission and fusion occur. The universe is expanding. Any account of science that fails to recognize that it was the *discovery* of these things that has driven the course of science more than any other factors just gets things massively and egregiously wrong. (See Wootton 2015: 57–109 on the nature, history, and importance of the concept of discovery.)

Crucially, many discoveries were completely unexpected; *nothing* in our background knowledge anticipated them. Nature is full of surprises, from finding a living *coelacanth*, to "jumping genes," to the discovery, in the first decade of this century, that not only is the universe expanding, the expansion is *accelerating*.[3] What is most relevant for scientists about such a discovery is not that the processes of its discovery were an exemplar of good science, though it may well have been. What matters most for the course of science is that, so far as scientists can tell, *it is so*.

Every such discovery points to the need for further discovery. DNA has a double helix structure. How, then, is the information in base pair sequences of DNA transferred from the cell nucleus, where the DNA is, to the cytoplasm where protein synthesis occurs? How does DNA mutate and how is damage to it repaired? How does DNA fold into chromosomes? The scientists making such investigations may indeed find the methods, ideas, and practices associated with an exemplar to be useful, but *what* they are striving so hard to find out is how nature can *be that way*. (See the long quote from molecular biologist Matthew Meselson in Morris 2018: 147–148)

My conclusion is conditional: IF the Kuhn of *Structure* is taken as defending relativism or irrationality, the arguments and analyses he offers there do not adequately support such claims. Kuhn's key ideas are too often nebulous, overstated, incoherent, or just wrong. In the next chapter, we try to get clear on the extent to which he was right.

Thomas Kuhn: Friend of Science?

4

Despite my many criticisms of *Structure* in the previous chapter, I do recognize that there were some things that Kuhn got right in these 13 sections. Not all of these insights are original with him, but they needed to be said as correctives to the claims made by earlier philosophers of science such as the logical positivists.

1. Past scientists should be understood and evaluated in their own historical context, and not faulted for failing to know what they could not have known.
2. Scientific reasoning is not rigidly rule-governed, but is guided by sets of mostly tacit assumptions about how science should be done and what constitutes good science. Scientific inquiry is not the application of algorithms, but the employment of rational argument and of practical skills learned and refined through scientific training.
3. There are revolutionary episodes in science that introduce radical discontinuities such that at the end of such revolutions, the field looks so different that hardly any ideas remain from its prerevolutionary aspect. (On the other hand, there may be considerable continuity of *methods*.) Scientists who live through such revolutions often feel that they and their fields have undergone a fundamental change of perspective.

4. Like everyone else, scientists are disposed to see only what they expect to see. Their commitment to theories can blind them to unexpected evidence. Conversely, once a new understanding is achieved, scientists will notice things to which they were previously oblivious.
5. There can be no pure observation language that reports raw, uninterpreted sensations. All such reports will involve an element of judgment or interpretation.

I consider these insights to be important but hardly deserving of the exalted status to which *Structure* has been elevated—and certainly unworthy of being treated as Holy Writ by the various academic science critics.

As noted earlier, *Structure* was far from Kuhn's last word. In his book *The Rationality of Science* (1981), W.H. Newton-Smith titles his chapter on Kuhn's "From Revolutionary to Social Democrat." In Newton-Smith's view, Kuhn softened his views over his career, evolving from the young firebrand to the older moderate. In his later writings, starting with the Postscript to *Structure*, and including essays published in the volumes *The Essential Tension* (1977) and *The Road Since Structure* (2000), Kuhn attempts to expand, clarify, and perhaps modify his distinctive claims.

In his "Postscript" to *Structure* written in 1969 in response to critics, Kuhn amplifies his ideas about incommensurability and adds comments that seem intended to placate the scientific rationalist. There he says this about debates over theory choice:

> . . . I have argued that the parties to such debates inevitably see differently certain of the experimental or observational situations to which both have recourse. Since the vocabularies in which they discuss such situations consist,

however, in the same terms, they must be attaching some of those terms to nature differently and their communication is inevitably only partial. As a result the superiority of one theory to another is something that cannot be proved in debate. Instead, I have insisted, each party must try, by persuasion, to convert the other.

(Kuhn 2012: 197)

Kuhn insists, however, that he never intended to imply that proponents of opposing theories cannot communicate at all or that persuasion is not by good reasons. He does deny that intertheoretic debates can be settled by anything resembling logical or mathematical proof (Kuhn 2012: 198). He continues:

Nothing about that relatively familiar thesis implies either that there are no good reasons for being persuaded or that those reasons are not ultimately decisive for the group. Nor does it even imply that the reasons for choice are different from those usually listed by philosophers of science: accuracy, simplicity, fruitfulness, and the like. What it should suggest, however, is that such reasons function as values and that they can thus be differently applied, individually and collectively, by men who concur in honoring them . . . There is no neutral algorithm for theory-choice, no systematic decision procedure which, properly applied, must lead each individual in the group in the same direction.

(Kuhn 2012: 198)

It is such passages in Kuhn's works that prompted Richard Bernstein in his enlightening book *Beyond Objectivism and Relativism* (1983) to construe Kuhn, not as an apologist for irrationalism, but as offering a view of scientific reasoning as very similar to what Aristotle called "practical reasoning" (*phrónēsis*) (Bernstein 1983: 54). For Aristotle, practical reasoning differs both from theoretical reasoning (*epistēmē*)—the kind or reasoning, for instance, employed in proving a mathematical theorem—and from *téchné*, the kind of reasoning used, for instance, by engineers in designing useful things.

Practical reasoning involves deliberation and choice. It is not algorithmic or rigidly rule-bound. The aim of practical reasoning is to realize certain general values within particular situations. Applied ethics employs practical reasoning. Everyone will agree on certain values such as justice but determining just how to implement justice in the complexities and contingencies of a particular circumstance can require much discussion and debate. For instance, which policies best promote racial justice-reparations, affirmative action, or what? The only rational way to settle these issues is to argue them out.

Likewise, on the model of scientific reasoning as practical reasoning, debates over theories would not be settled by automatic decision procedures, but by rational deliberation between the qualified parties. The debate will be guided, but not determined, by shared scientific values and standards with disagreement over which theory best exemplifies the virtues endorsed by those standards and values. That is, which theory best exemplifies such theoretical virtues as simplicity, accuracy, consistency, scope, and fruitfulness (Kuhn 1977: 321–322). All scientists agree that good theories should have these virtues. The debate will be over which of the theories under consideration is the simplest, most accurate, broadest in scope, etc. Such debates are eminently *rational*, and a good scientist will be one that is skilled in the deliberations and debates over how best to instantiate those values in situations of theory choice. That such debates persuade but do not coerce agreement in no way detracts from their rationality.

By "persuasion" here Kuhn clearly does not mean the sort of "persuasion" he meant in Structure, where he said that proponents of opposing paradigms were forced to use circular argument and propagandistic techniques of mass persuasion (Kuhn 2012: 94). Politicians, ideologues, and advertisers—all those who stand to gain by manipulating your beliefs—freely use any means fair or foul to get you to accept their claims. They might use reason when that suits their aim, but they freely resort to *ad hominem* abuse, loaded questions, equivocation, big lies, skewed statistics, and the whole litany of fallacies and subterfuges used by those with manipulative intent. Here, Kuhn seems to be saying that scientists have a higher standard and support their claims by appealing to *rational* reasons in the form of recognized theoretical virtues such as scope, simplicity, etc.

For instance, Darwinian evolution evinces a number of such theoretical virtues. The theory is simple, postulating a process of natural selection that must occur given certain well-known facts about inheritance, variation, and natural environments. It has extremely broad scope, explaining a wide variety of different natural facts, from anatomy, to the geographic distribution of organisms, to the fossil record, to embryology. In fact, it has been rightly said that *nothing* in biology makes sense except in the light of evolution. Further, evolutionary theory has certainly proven one of the most fruitful in the history of science, making understandable the history, diversity, and distribution of life on Earth.

Bernstein's interpretation of Kuhn is plausible, particularly with respect to some of Kuhn's later writings, such as his essay "Objectivity, Value Judgment, and Theory Choice" (Kuhn 1977: 320–339). For the scientific rationalist, it is also an attractive interpretation, since Kuhn can be seen as a friend rather than an antagonist.

A second notable effort to bring Kuhn into the rationalist camp was made by Wesley Salmon, one of the leading philosophers of science of the twentieth century. In his essay "Rationality and Objectivity in Science," Salmon argues that Kuhn's view of rational scientific debate over theoretical virtues may be plausibly interpreted as similar to features of the Bayesian approach to scientific rationality (Salmon 2005: 93–116).

Bayes' Theorem is a theorem of the probability calculus. Without getting into the mathematical details, the theorem can be employed to specify the factors involved in the evaluation of a theory on the basis of evidence. That is, it tells us what we have to consider when determining how strongly a theory is supported by particular evidence. Essentially three factors are involved: (1) How probable is the theory on the basis of our background knowledge, that is, all of our relevant knowledge *except* the particular evidence in question? (2) How likely is it that we would have that particular evidence given that the theory is *true*? (3) How likely is it that we would have that particular evidence *whether or not* the theory is true, that is, given the *total probability* of the evidence? Bayesians are those who think that scientific rationality, how scientists confirm theories, can be modeled on Bayes' Theorem. Here I will not offer an evaluation either for or against the Bayesian model of scientific rationality, but will merely note that it has broad support among philosophers of science.

Consider, for instance, the degree to which the discovery of feathered dinosaurs supports the theory that birds descended from dinosaurs (much more on this in the sixth chapter). According to the Bayesian model, paleontologists, at least implicitly and informally, consider three sorts of questions in making an evaluation: (1) On the basis of our background knowledge, how probable is it that birds descended from dinosaurs? (2) How likely were there to be feathered dinosaurs if birds *did* descend from dinosaurs? (3) How likely were there to be feathered dinosaurs *whether or not* birds are the evolutionary descendants of dinosaurs? If, (1) given our background knowledge, it is not improbable that birds descended from dinosaurs, and if (2) it is quite likely that some dinosaurs would have feathers if birds did evolve from dinosaurs, and if (3) it is not very likely that some dinosaurs would have feathers if birds did not descend from dinosaurs, then, say the Bayesians, feathered dinosaurs are confirming evidence for the theory of dinosaur/bird evolution.

Salmon considers Kuhn's criterion of consistency. Consistency can be taken in two ways, as the internal consistency of a theory—whether any of its elements contradict—and external consistency, or compatibility with accepted theories (Salmon 2005: 115). Salmon takes consistency in these two senses to apply to the evaluation of the background probability of a proposed theory. Obviously, if a theory is seen to be internally inconsistent, so that some parts of it contradict other parts, then it is a non-starter and is not considered until and unless the contradictions are worked out. If a new theory is compatible with accepted background theories, this increases the plausibility of the new theory, that is, we will consider its background probability to be enhanced. If, on the other hand, it contradicts those established theories, the new theory loses credibility and we think it less probable. To take an example not mentioned by Salmon, if a newly proposed theory is found to contradict the laws of thermodynamics, this would pretty much discredit that theory.

Salmon also considers Kuhn's criterion of simplicity. In the physical sciences, the simplicity of a proposed theory vis-à-vis its competitors is taken as conferring greater plausibility on that theory, that is, other things being equal, scientists will regard it as having higher prior probability than its competitors. (As Salmon notes, p. 101, in the social/behavioral sciences, simple theories are often too simple.) Kuhn's theoretical virtues of consistency and simplicity may therefore be seen as factors the Bayesian will consider in estimating the prior probabilities of theories:

> Kuhn's criteria of consistency (broadly construed) and simplicity seem clearly to pertain to the assessments of the prior probabilities of theories. They cry out for a Bayesian interpretation.
>
> (Salmon 2005: 113)

Salmon also finds a connection with respect to Kuhn's criterion of fruitfulness (Salmon 2005: 113). One interpretation is that a fruitful theory is one fertile in predicting previously unknown phenomena (Salmon 2005: 113). For instance, special relativity predicts the phenomenon of time dilation, that is that clocks slow down on a body that has been accelerated as compared to one that has remained stationary. Such a phenomenon was unknown prior to the formulation of special relativity, and would have been very unexpected.

Bayesians call the probability of a piece of evidence given background knowledge the "expectedness" of that evidence. The more unexpected a piece of evidence, that is, the more surprising it is, the more strongly it confirms the theory that predicts it. When time dilation was experimentally measured, it strongly confirmed special relativity because it was highly unexpected otherwise. Here, then, we appear to have another connection between Kuhn and the Bayesians. Fruitful theories (on one interpretation) will make bold and surprising predictions, and Bayesians can explain just why these predictions are particularly potent confirmations of the theories that predict them.

Bernstein and Salmon therefore offer plausible proposals for placing Kuhn—at least the later, post-*Structure* Kuhn—in the rationalist camp. However, I cannot fully accept such irenic resolutions. Two issues remain: Kuhn's "mature" views of incommensurability and his antirealism, neither of which, it seems to me, can be acceptable for the rationalist.

Muhammad Ali Khalidi identifies what he considers Kuhn's "mature" understanding of incommensurability (Khalidi 2000: 173–175). The view that Kuhn eventually settled on, says Khalidi, equates incommensurability with untranslatability between different theories. To say that two theories are incommensurable means that the translator cannot render a sentence from one in the terms of the other without some loss of meaning and the consequent danger that the truth value of the original sentence will be lost in the translation. There are two specific problems here. The first we might call the "holism" problem and the second the "disparity" problem.

For Kuhn, the meaning of the terms of a theory can only be understood holistically, that is, in the context of that theory. This is because those terms are *inter-defined* with other terms in the theory and therefore the set of interconnected terms has to be grasped as a whole. The organic connections of meaning between the terms of a theory are broken if those terms are appropriated and put into the radically different conceptual environment of another theory. The consequence is that a cluster of concepts that is understood as an interconnected whole in one language is broken up into different and inequivalent concepts when those terms are explicated in another language.

The second sort of problem identified is an intractable disparity between the meaning of concepts as expressed in one language and how another language attempts to construe those concepts (Khalidi 2000: 175). A term in one language may have no precise equivalent in another language. The term may be rendered by clumsy paraphrases, but these paraphrases break up what was a unitary concept in the original language to a cluster of concepts in the translation. Allegedly, then, the sense in which a term stands for a unified concept in the original language will be lost in the translation.

Khalidi spells out the alleged consequences of these two issues:

> Therefore, according to Kuhn's mature view, it is not possible to phrase all of the claims of two scientific theories in a single language so that they can be put side by side and their exact points of difference pinpointed. Kuhn therefore denies the possibility of what is perhaps the most direct and natural method of comparing two scientific theories. As a result, choices between theories are not based on a point-by-point comparison.
>
> (Khalidi 2000: 175)

In my pedagogical experience, I have encountered both of the above sorts of issues. In teaching Aristotle's *Nicomachean Ethics*, I have found that the term *"eudaimonía"* cannot be translated by any single word or phrase in English. Usually, it is translated as "happiness," which, for various reasons, is unfortunate. "Thriving," "well-being," and "flourishing" are closer approximations, but still not entirely satisfactory. Further, to understand *eudaimonía* as Aristotle intended it, you have to understand the concept in conjunction with other key terms such as virtue (*aretē*) and the human function (*érgon*). So, what for Aristotle was a single concept, *eudaimonía*, has to be explicated in English with reference to a cluster of concepts. However, there is no reason to see this as posing an insurmountable problem. Any complex concept can be analyzed into elements, and then put back together and understood as a single concept. That is just what we mean by "understanding" a complex concept.

Conveying Aristotle's meaning to an audience of mostly tepidly interested undergraduates is a challenge, but there is no reason to give up on such a task. Neither the fact that there is no single word or phrase that precisely translates *"eudaimonía"* nor the fact that its meaning is organically interconnected with other Aristotelian terms presents an insurmountable impediment to understanding. Word-for-word translation is not the only tool we have in understanding thoughts expressed in another language and transmitted from a very different cultural context. There is also exposition, that is, the explanation, at whatever the required length, of the full richness of the meaning and significance of terms. We can engage in the kind of exposition and analysis Kuhn performs in his historical works on the Copernican Revolution and the beginnings of quantum physics.

Once, then, we have grasped Aristotle's meaning adequately (and Kuhn gives no reason to think that we cannot), we can then *rationally* assess his claims, in part by comparison—even point-by-point comparison—with very different ethical theories such as Kant's. In other words, there is *no* incommensurability between Aristotle's and Kant's ethical theories. They are comparable and can be judged *vis-à-vis* one another.

The language of particular theories is therefore not the only language available to scientists. I take it as true, indeed trivially true, that the concepts of one theory often cannot be expressed in terms of the concepts of an opposed theory. Clearly, for instance, many concepts of special creation cannot be expressed in the terms of evolutionary theory. Evolutionary theory contains nothing corresponding to supernatural speech-acts of divine creation ("Let there be ..."). However, this does not mean that evolutionists cannot *completely* understand the tenets of special creation. Quite a few do, and criticize it cogently (see, e.g., Kitcher 1982, and 2009).

If Kuhn were to insist that *something* is always lost in translation, then we have to ask whether, in any particular instance, that "something" is specifiable. If it is, then we can express it and it is not lost. If it is inexpressible, then I would have to conclude that Kuhn's assertion of non-translatability just reports his own intuition. I have other intuitions.

Is point-by-point comparison of theories impossible? Not at all. Scientists do it all the time, even proponents of very different theories and even in revolutionary episodes. Ptolemaic theory predicts that Venus does not show phases; Copernican theory predicts that it does. It does. Therefore, Copernican theory successfully makes a prediction and Ptolemaic theory fails. The only way to deny the success of Copernican theory and the failure of Ptolemaic theory on this point would be to say that the observation terms "Venus" and "phase" had changed radically in meaning and reference between the two theories. As noted in the previous chapter, this is implausible in the extreme.

Examples of such point-by-point comparability between very different theories may be multiplied at will. Galileo's

Siderius Nuncius and Darwin's Origin, for instance, are full of such comparisons, fully comprehensible (and hence rationally debatable) by friend and foe alike. Kuhn therefore seems once again to have badly overstated things, and to have equally badly underestimated the human capacity to communicate and make rational judgments, even when comparing radically different theories.

At this point, I will venture the opinion that incommensurability of meaning does not exist. I regard it as a pernicious philosophical canard that we would all do better without.

Finally, there is the question of Kuhn's antirealism. Kuhn himself sometimes seemed unsure whether to call himself a realist or not (Marcum 2015: 140). However, Richard Rorty has no doubts:

> Kuhn turned away from philosophy to history for a time, while preparing a history of the origins of quantum mechanics . . . But since the publication of that book the bulk of his work consisted in detailed defense of the claim that there is no language-independent reality, no single "Way that the World Is. . . ."
>
> (Rorty 2000: 204–205)

Rorty then quotes Kuhn:

> . . . I aim to deny all meaning to claims that successive scientific beliefs become more and more probable or better and better approximations of to the truth and simultaneously to suggest that the subject of truth claims cannot be a relation between beliefs and putatively mind-independent or "external" world.
>
> (Kuhn 1993: 330)

Antirealism as Rorty understands it, and the sense in which he imputes it to Kuhn, seems to be this: When two competing theories make conflicting claims, there is no fact of the matter that could make one right and the other wrong. There is no mind-independent external world with intrinsic properties that could make one theory true and not another.

If this kind of antirealism was Kuhn's view, it would seem to make sense of much of what he says in Structure. Pendulums and the moons of Jupiter were not discovered, because that would mean that we would know something about the mind-independent, external world, namely that it contains pendulums and moons of Jupiter. Pendulums and the moons of Jupiter therefore did not exist until Galileo "saw" them. Kuhn's rhetoric (e.g., "world changes")—which even he seemed at times to struggle to understand—makes sense in the light of such antirealism.

Further, if science is not an effort to understand the intrinsic nature of external physical reality, then science must be some sort of construct. That is the only alternative. Science will not be about the natural world. There is no natural world, only our ideas about the "world." Nature is made, not discovered. Different paradigms really do create different worlds because those are the only worlds there are.

Alexander Bird, on the other hand, does not see Kuhn as an antirealist. Bird says that Kuhn holds that there is a mind-independent world, but regards that world as unknowable, or nearly so (Bird 2000: 137). For Kuhn, the world is like Immanuel Kant's "things-in-themselves (Dinge an sich)." For Kant, the world as we experience it is not the world as it is in itself, but the world as it must appear to beings with intellectual and perceptual faculties such as ours. The things-in-themselves are unknowable, yet, for Kant, they must be posited to account for the existence of the world we can know. Unlike Kant, Kuhn holds that we can say some very basic things about the natural world, such as that it must be differentiated, and not homogeneous (Bird 2000: 127). Bird continues:

> The reason we can say such things is that Kuhn believes that there must be such a world in order to account for the fact that adherents of the same paradigm by and large share the same experiences. This shows we can say more. The world-in-itself is causally responsible for our experiences (in conjunction with the paradigms instilled in our minds), even though we cannot say how.
> (Bird 2000: 127–128)

If Bird is right, then, strictly speaking, Kuhn is not an antirealist, and the world is not (purely) a construct. For Kuhn, the world-in-itself, like Kant's things-in-themselves, is the anchor to objective reality that keeps him from drifting into pure idealism or relativism. Yet, as the history of philosophy shows, things-in-themselves are fragile creations. They tend to get swept away by later and bolder thinkers who have no problem with biting bullets. Thus, Hegel dismissed Kant's things-in-themselves and frankly advocated idealism. Likewise, the social constructivists and postmodernists like Rorty saw no need to keep one toe planted in a putatively objective world, and enthusiastically embraced antirealism and relativism.

If, then, Kuhn was a realist, it was a kind of realism that asymptotically approached antirealism. Such "realism" is vaporous and insubstantial, easily blown away by the breezes of idealism or relativism. In my view, then, if Kuhn was not himself the prophet of irrationalism, he was, however unwillingly, its John the Baptist, making straight the path to it.

CODA: HISTORY, WHIGGERY, AND PROGRESS

We noted that Kuhn holds that, by definition, science progresses (Kuhn 2012: 169–179). However, he does not hold that science progresses in the sense of getting a clearer and more accurate representation of an objective natural reality. Rather, the progress is more like biological evolution in which succeeding populations of organisms are better than their ancestors at meeting challenges posed by the ambient environment. By analogy, succeeding scientific theories are better than earlier theories at addressing anomalies and solving problems posed by the intellectual environment, but this does not imply that they, in any absolute sense, mirror nature any better than the old theories. Our science today is far more complex and sophisticated than Aristotle's, and has far more problem-solving power, but, Kuhn holds, this does not mean that we grasp the putative essential nature of the world any better than Aristotle.

Other than a prior philosophical commitment to a Rortyan or Kuhnian antirealism, what would motivate the reluctance to say that there are some things that past scientists got wrong but that, so far as we can tell, we have gotten right? In other words, why should we doubt that science progresses in the straightforward sense that over time we come to know more and more about the natural world? One basic motivation for such reluctance seems to be a misunderstanding or misapplication of the concept of Whig history. To say that past scientists were wrong and that we now know better perhaps seems to be a whiggish imposition of current perspectives on a past intellectual milieu that is unfairly judged by standards that could not have been appreciated at that time. On the contrary, however, it is not only possible but necessary to understand the past in terms of what we now know. It cannot be ahistorical to tell the truth.

Consider once again the N-ray case examined in the second chapter. Suppose that a historian were asked whether Blondlot failed to detect N-rays, and, fearful of committing whiggish impropriety, replied that answering such questions is no part of the historian's business. All the historian can do is to present how things appeared to people *at that time*, and judgments about the rightness or wrongness of those perceptions are not within the historian's purview.

However, such a rule, if consistently applied, would forbid historians from saying that Napoleon lost the Battle of Waterloo. All they can do is say why it seemed that way to people at that time. However, if the perceptions of people at the time that Napoleon had lost still seem accurate to us today, we do not hesitate to say that Napoleon lost. Likewise, if the evidence of Blondlot's failure that convinced the physics community of the day still looks like good evidence to us today, there can be no reason not to say that, in fact, Blondlot failed to detect N-rays. Surely, we should seek to understand the past, as we attempt to understand anything, in terms of *everything* that we think we know. To choose otherwise is willful blindness, a dereliction of intellectual duty.

In general, there is nothing pejoratively ahistorical about judging that a past scientist was right or wrong. Of course, we must remember that a scientist who was wrong was not necessarily a bad scientist. We must not divide past scientists up into "good guys" and "bad guys" based on how well they anticipated current theories. Opposition, even tenacious opposition, to an ultimately successful theory need not imply any incompetence or bias on the part of the objector. Our sense of justice and fair play is rightly offended when Whig historians traduce a scientist, as they did Richard Owen, by ignoring the historical context and condemning him for not making judgments that are "obvious" only in hindsight. Yet to guard against such derogatory and unfair judgments, we do not need a blanket rule proscribing assessments of rightness or wrongness. We merely need to respect the philosophical commonplace that there is a difference between being reasonable and being right. Sometimes people are reasonably wrong and sometimes unreasonably right.

Still, we might be asked, by whose standards are we to judge that a past scientist or theory was a success or failure? Without blush or hesitation, we answer "by ours, of course!" Unless we espouse a relativism that regards all epistemic standards as merely parochial, and science as just another narrative, then we will see nothing objectionable about assessing scientific success or failure in terms of what, by our best lights, we now think to be so. To refuse to make such judgments based on our current scientific understanding is to say that the history of science must be conducted within a self-imposed cone of silence, where historians are unable to hear what, for everyone else, is common knowledge. Why would historians choose to operate under such a handicap in deference to a baseless methodological stricture?

Can We Have Good Science and the Right Values?

5

Is science the tool white males use to maintain their privileges over women and people of color? Are scientists aggressive ideologues who marginalize religious people by imposing a dogmatic naturalism? Critics on both the left and the right have regarded science and scientists as a threat to their basic values. Marxists have seen science as a tool of capitalism that heedlessly maximizes profits at the expense of people. Environmentalists decry the development by the chemical industry of profitable pesticides such as neonicotinoids, substances they blame for various deleterious environmental effects. Conservative Christians, on the other hand, such as lawyer/activist Phillip Johnson have charged that science has abandoned impartiality and objectivity by arbitrarily excluding any hypotheses of creation or intelligent design (see Johnson 1991).

Science has always threatened social, religious, and moral values, sometimes to the extent of arousing bitter controversy and conflict. In the fifth century BCE, Anaxagoras outraged his fellow Athenians by claiming that the sun is not a god but a big hot rock. As is well known, Galileo provoked the Church hierarchy to haul him before the Inquisition, force his recantation, and sentence him to house arrest for the remainder of his life. Equally famously, Darwin outraged conservative proper Victorians. There is an amusing anecdote about the Anglican bishop's wife's reaction upon hearing of Darwin's theory: "Oh!

DOI: 10.4324/9781003105817-6

My dear! Descended from apes! Let us hope that it is not true, and, if true, that it does not become widely known!" And it is not just the staid and the stodgy who are susceptible to outrage. In the 1970s, left-wing activists reacted furiously and imputed racism and sexism to E.O. Wilson's theories of sociobiology. So, science that is exhilarating to some is repugnant to the deeply entrenched values of others.

The trustworthiness of science, which is the topic of this book, comes down to its cognitive credentials. If well-confirmed scientific theories are the most trustworthy accounts we have, then, to the extent that we value rationality, we have at least a strong prima facie duty to act on those theories and not on others. For instance, we should teach evolutionary theory rather than creationism in the public schools. Yet, anti-science animus, stemming from either the left or the right, is certainly largely due to the perception that science threatens basic moral, social, and political values. To some extent this is inevitable. As it says on a T-shirt I frequently wear, "science does not care what you believe." Put in a less confrontational way, you cannot expect that the results of empirical inquiry will support your favored creeds or convictions. If you believe that the universe was created in six literal days 6000 years ago, or that vaccines cause autism, or that human-caused climate change is a hoax, then you are wrong.

In this chapter, I will try to sort out some of the issues of science's relation to moral and social values. Everybody agrees that science is and should be guided by "epistemic" values, that is, such values as objectivity, consistency, testability, etc., that bear upon the truth or rationality of scientific claims. What, though, about "non-epistemic" values such as moral, social, or political values? Can such values be injected into scientific practice without corrupting scientific objectivity? Would a Christian science, a feminist science, or a Marxist science still be science? Might some values such as diversity actually enhance the objectivity of science by working to exclude partial, one-sided, or biased judgments? On the other hand, does the disparagement and undercutting of science by ax-grinders of the left and right threaten to diminish the status of science in society? Should scientists themselves draw upon moral and social values in making decisions about which hypotheses to accept?

I will begin by examining the *wrong* way to mix science and values—Sandra Harding's feminist standpoint theory. I argue that Harding's claim to propose a stronger form of scientific objectivity—by putting the *right* politics into science—is wrongheaded, and that her approach would reduce science to just another ideology. In the next section I consider much more creditable discussions of values in science by Heather Douglas.

GETTING THE *RIGHT* POLITICS INTO SCIENCE

There is a well-known photograph of the fifth Solvay Conference held in Brussels in October 1927, subsidized by wealthy Belgian industrialist Ernst Solvay. The conference gathered the most renowned physicists, including many past and future Nobel Prize winners, to discuss the newly developed quantum mechanics. Of the 29 persons posing for the picture, only one is a woman, Marie Curie. Perhaps she should count for two since she won Nobels in both physics and chemistry. Today, nearly a hundred years since that photograph was taken, things have improved, but women remain underrepresented in physics and astronomy, as the American Institute of Physics reported in 2019:

> The participation of women in physics has greatly increased since the 1920s . . . however, the proportion

of women among physics students and faculty members is still below that of other disciplines. Women earn over 50% of all bachelor's degrees . . . but women earn only 21% of physics and 33% of astronomy bachelors' degrees. https://www.aip.org/statistics/reports/women-physics-and-astronomy-2019

The reasons that women remain underrepresented in some STEM fields are complex and we cannot pursue them here. Surely, though, it is uncontroversial that girls and young women should be strongly encouraged to pursue and develop their interests in science, math, and engineering. Making these fields welcoming to women is no more than basic equity. Further, STEM studies are so important that it is essential to have no barriers to participation by the best minds whomever they might be.

For some of the more radical feminists of the science wars era, issues such as inclusion, diversity, and access barely scratched the surface of the real problem about science and women. As they saw it, the deep problem was that science itself had been so thoroughly corrupted by sexism that its fundamental assumptions had to be overthrown, particularly its ideas about objectivity. "Objectivity is what a man calls his subjectivity" is the slogan I once heard a young woman proclaim at a philosophical gathering. What I think she meant was that scientists—male scientists—had enshrined a concept of objectivity as disinterested, neutral, and impartial. However, this appearance of objectivity is a smokescreen for an ideology that automatically favors the privileged and elite—white males—and marginalizes everyone else. It is an insidious subterfuge to hide one's vested interests behind a façade of disinterested inquiry. Science is a product of Western, linear, and male (i.e., bad) ways of thinking, and its logic is a tool of oppression.

One of the best-known feminist writers making such a claim about objectivity is Sandra Harding. Let me hasten to add that Harding is not presented here as representative of feminist philosophers of science, many of whom have expressed strong disagreement with Harding. I focus on Harding here because space is severely limited, and she expresses a particularly challenging set of claims that any scientific rationalist will have to consider. Here we consider her views as stated in her 1991 book, *Whose Science? Whose Knowledge? Thinking from Women's Lives*.

Harding defends what she calls "feminist standpoint theory." The fundamental assumption of such theory is that knowledge is "socially situated," that is, that the specific social organization of a society structures knowledge and sets limits on what can be known. In particular, when a society is organized with rigid hierarchies, with one group dominant and another oppressed, these differences have cognitive consequences, making insights available to some groups and distorting the beliefs of others. Harding holds that the advantage lies with the oppressed. It is often said, for instance, that slaves understand their masters in minute detail; their survival depends on it. Masters, however, know their slaves only superficially. Harding thinks that the marginalized position of women makes it possible for them to see things clearly when their advantages make men oblivious. Therefore women's experience, correctly interpreted, can provide a basis for richer and deeper understanding, in the sciences as well as in other fields. As Harding puts it:

> Knowledge emerges for the oppressed through the struggles they wage against their oppressors. It is because women have struggled against male supremacy that research starting from their lives can be made to yield up clearer and

more nearly complete visions of social reality than are available only from men's side of these struggles.

(Harding 1991: 126)

The key insight necessary to begin the process of incorporating information from women's lives is to recognize that all science is inherently and irremediably political and value-laden. Harding calls the ideal of "value free, impartial, and dispassionate" (Harding 1991: 138) objectivity "weak" objectivity (Harding 1991: 143–149). Proponents of such "weak" objectivity seek to free knowledge from all human constraints and permit nature alone to determine our views and thereby to lead us to know it as it really is. Harding cites Thomas Kuhn, Steven Jay Gould, and many of the sociological and social constructivist studies of science to reject such an idea as a pernicious myth. She claims that science as pursued by the dominant white males of the Western world has again and again structured their science to promote their hegemony and to exclude and marginalize the scientific and epistemological stances of "Others," such as women, African Americans, and those of Third World descent (Harding 1991: 140). The appeal to impartiality and objectivity is a stratagem to disguise and legitimize such marginalization, because the only values and interests eliminated are those of the "Others," and not the dominant "Western, bourgeois, and patriarchal" values (Harding 1991: 145).

The cure for "weak" objectivity is not relativism, says Harding, but to adopt "strong" objectivity, a kind of objectivity that will be less "partial and distorting" than "weak" objectivity (Harding 1991: 143). The first step towards a stronger and less distorting ideal of objectivity is to realize that political and social interests are not "add-ons" that better methods can eliminate from science; they are essential elements (Harding 1991: 145). Clearly, then, for Harding, if all science is inevitably value-laden, then it is essential that it be laden with the right values.

To achieve political rectitude, science must incorporate women's experience as an essential element. However, it is not just the experience of women *per se*. After all, women's experience has itself been warped by patriarchy and misogyny. For instance, women had to learn to define as rape sexual assaults that occur in marriage (Harding 1991: 123). Thus, it is not the raw data of women's experience that is crucial, but women's experience as interpreted by feminists. The upshot is that Harding thinks that the way to improve science is for it to adopt a specifically feminist standpoint and set of values.

Unsurprisingly, Harding's claims drew a number of pointed rebuttals. Cassandra Pinnick criticizes feminist epistemologies such as Harding's by noting that they fail to resolve an inherent tension:

> . . . any feminist epistemology which radically challenges traditional theories of knowledge is unable to resolve the tension between (a) its thesis that every epistemology is a sociopolitical artifact, and (b) its stated aim to articulate an epistemology that can be *justified* as better than its rivals.
>
> (Pinnick 1999: 295)

As noted, Harding thinks that her "strong" objectivity would be less "partial and distorting" than the "weak objectivity" that hides male privilege behind an appearance of neutrality and impartiality. Yet to make reasonable judgments about partiality requires that one have an *impartial* criterion to distinguish between those standpoints that are partial and those that are not. If *all* standpoints are sociopolitical artifacts, then so is one's own. To judge that one's own standpoint is superior on the basis of criteria grounded in that very standpoint is obviously viciously circular. Harding's project therefore appears self-refuting; it requires impartiality while denying that it exists.

Pinnick disputes Harding's claims that feminists, because of their allegedly marginalized status, are better qualified to make impartial and accurate judgments. Harding offers no evidence or data to indicate that feminists have had a better track record of scientific achievement than nonfeminist researchers:

> Specifically, Harding needs to show that politically motivated research, under the guidance of feminists, accomplishes scientific aims better than research done under the auspices of the traditional empiricist, socially and politically disengaged, ideal inquirer.
>
> (Pinnick 1999: 302)

Really, if, as Harding maintains, *all* epistemologies are tools of vested interests, then no epistemology, including feminist standpoint epistemology, can claim to be more objective and less partial and distorting (Pinnick 1999: 303). Finally, and amusingly, Pinnick asks, if the marginalized status of feminists gives them an epistemological advantage, what happens to that advantage if they succeed in their goal of achieving political equality? (Pinnick 1999: 303)

In her critique of Harding, Ellen R. Klein charges that Harding has waffled on her definition of "objectivity" (Klein 1996: 39). Sometimes she seems to mean no more than that objectivity corresponds to consensus within scientific communities, but Klein sees nothing particularly feminist in that definition (Klein 1996: 39). If Harding means that science has to be based upon an explicitly feminist agenda, then such a political program will, at best, just be irrelevant to most scientific inquiries, and, at worst, would be no less distorting than the male biases purportedly entrenched in science (Klein 1996: 39).

To support Klein on this point, I would note that for many feminists it is a tenet of ideological faith that gender is a social construct, i.e., that while sex is natural, behavioral differences between men and women are entirely socially constructed, and in a society engineered around feminist values stereotypical "masculine" and "feminine" behaviors would fade away. But what if they are wrong? What if, for instance, as Steven Pinker argues in his 2002 book *The Blank Slate*, and Debra Soh in *The End of Gender* (2020), there are innate dispositional differences in the behavior of men and women? This is an empirical question, and *any* ideological demand for a particular outcome, even a "liberating" one, would seem to be "partial and distorting." As noted earlier, you cannot demand that the results of empirical inquiry conform to your doctrinal requirements.

Klein asks what, really, is the idea of objectivity that has been shared by scientists and most philosophers of science? Klein notes that being value-free is not essential to science or scientific method (Klein 1996: 40–41). Being disinterested is not the same thing as being uninterested. Scientists are clearly going to be vitally interested in their claims, but seek methods that permit their results to be disinterested, i.e., unbiased (Klein 1996: 40). When feminists claim that no science is unbiased, Klein says that, ironically, they appeal to a male authority, Thomas Kuhn (Klein 1996: 44). Feminists cannot just uncritically invoke Kuhn to justify their dismissal of scientific objectivity (Klein 1996: 45).

In her recent book *Why Trust Science?* Naomi Oreskes offers a defense of Harding. She takes Harding to be a friend of science who wants to enhance objectivity by increasing the diversity of participants in the scientific process:

> Harding mobilized the concept of *standpoint epistemology*— the idea that how we view matters depends to a great extent on our social position . . . to argue that a greater diversity would make science stronger. Our personal experiences—of wealth or poverty, privilege or disadvantage, maleness or femaleness, heteronormativity or queerness, disability or able-bodiedness—cannot but influence our perspectives on and interpretation of the world. Therefore, *ceteris paribus,* a more diverse group will bring to bear more perspectives on an issue than a less diverse one.
>
> (Oreskes 2019: 50; emphasis in original)

In short, ". . . the best way to develop objective knowledge is to increase the diversity of knowledge-seeking communities" (Oreskes 2019: 51).

So, Oreskes interprets Harding as making the same sort of claim as an advertisement by the firm Genentech in *The Atlantic Monthly:*

> Science demands diversity. After more than 40 years of tackling the toughest medical challenges, we know that approaching any problem from a single point of view is setting a course for failure. Success depends upon

welcoming diverse approaches, challenging the status quo and exploring hypotheses from all angles. Science demands diversity and so do we.

(*Atlantic Monthly*, October 2020: 71)

It is intuitive that a greater diversity of participants in any cognitive enterprise will increase the variety of perspectives and that a heterogeneity of viewpoints could enhance objectivity as the biases of one perspective are corrected by the insights of others. However, it is important to ask what kinds of diversity will be most helpful in which sciences. For instance, would religious or political diversity matter more or less than gender diversity or differences in sexual orientation? What kinds of diversity would enhance objectivity in, say, astrophysics as opposed to, say, primatology? To reply that we should enhance *all* kinds of diversity is not much help because people are diverse in innumerable ways.

The objectivity of science is largely a matter of the rigor of the methods and techniques employed, and it is not clear whether the factors mentioned by Oreskes—wealth, poverty, maleness, femaleness, etc.—would have much bearing on the development of stringent methods. Henrietta Leavitt discovered the period/luminosity relationship of the Cepheid variables, but her discovery would seem to have had much more to do with her intelligence, diligence, and visual acuity rather than her femaleness.

More basically, I think that Oreskes has not recognized the genuinely radical nature of Harding's arguments. Harding is making a much stronger claim than the rather mild assertion that greater diversity will provide a richer set of perspectives. She is not merely making a plea for inclusion and equity. She is not calling for a diversity of perspectives. On the contrary, she is calling for the dominance of one perspective—hers. Harding is an ideologue. She is just as much an ideologue as those who would like for science to be based upon fundamentalist Christian or Marxist/Leninist principles. Remember, she thinks that sexism has corrupted science *through-and-through*, and that the political nature of science is not eliminable. Therefore, since all science is inevitably political science, it must be thoroughly reconstituted based on the *right* politics, as determined by feminists of her stripe.

A defender of Harding might say that other ideologies are oppressive, but that feminism is liberating. Oppressive ideologies distort, but liberating ideologies elucidate. However, *every* ideology sees itself as liberating. Every ideology says "Ye shall know the truth, and the truth shall make you free. And *we* have the truth." If Harding gets to try to base science on her ideology, she gives tacit permission for Christian fundamentalists to do the same thing. It then becomes a power play to see whose ideology wins. (In a battle between feminists and fundamentalists, my money would be on the fundamentalists, since they are more numerous, better funded, better organized, and fight dirtier.)

So, before we can accept Harding's total overhaul of science, some questions must be answered; (a) Apart from appeals to authority, has Harding shown that science cannot be done apolitically? (b) Has she shown that science in general would be done better if done from her standpoint? What would, say, astrophysics or molecular biology look like if done from the feminist standpoint? (c) Given that *some* sciences (particularly medical and social sciences) might need to "think from women's lives," who gets to speak for women? Women are a very heterogeneous group. Can Harding speak for all of them?

When science serves an ideology—any ideology—it stops being science. Disinterested objectivity is the only kind of objectivity. Psychologist Stuart Ritchie puts it like this:

> To my mind, though, calling for scientists, or science in general, to take on any set of socio-political commitments is unwise, even if it might help solve specific problems in the short term. We should instead do our best to limit the effects of our own prejudices on our scientific decisions and analyses . . . Trying to correct for bias in science by injecting an equal and opposite dose of bias only compounds the problem and potentially invites a vicious circle of ever-increasing division between different ideological camps.
>
> (Ritchie 2020: 118–119)

In a somewhat more astringent tone, Susan Haack states what should serve as the quietus to all attempts such as Harding's:

> Other feminist critics, holding that science is unavoidably a political enterprise, and thus far informed by masculinist values, advocate its transformation by the injection of more progressive, feminist values; as if we hadn't learned from the awful examples of Nazi and Soviet science that to put pressure on scientists to arrive at politically desired conclusions is to court disaster.
>
> (Haack 2003: 315)

WHAT IS THE SCIENTIFIC VALUE OF VALUES?

A much more insightful analysis of science and values is offered by philosopher Heather E. Douglas in her book *Science, Policy, and the Value-Free Ideal* (2009), and in the Descartes Lectures delivered at Tilburg University in the Netherlands in 2016 and published as *The Rightful Place of Science: Science, Values, and Democracy* (2021).[1]

Let me begin by making clear that there is no explicit or implicit anti-science agenda in Douglas's analysis. On the contrary, her aim is to increase confidence in science. She strongly supports the epistemic integrity of science against the demands of ideology and profit:

> Social and ethical values can have a distorting effect on science, as evidence by the cases of sexist science uncovered by feminists. Such cases are just one way in which social and ethical values can distort science. Occurrences of manufactured doubt show the influence of social ideologies on scientific research. Because the purveyors of doubt care so much about unfettered capitalism [See Chapter Six on the topic of climate change.], they are willing to distort the scientific record to forestall unwelcome policies . . . And some cases of scientific fraud can be viewed as a pernicious influence of social values, when scientists are so sure of how the world should be, they make up the data to show that it is that way . . . Social and political values also drove such catastrophic cases as the influence of Trofim Lysenko on Soviet science under Stalin.
>
> (Douglas 2021: 22)

Douglas later clarifies that, while it is unacceptable to reject scientific evidence when it conflicts with values, one is justified in placing a heavier burden of proof on claims that so conflict:

> New evidence should always be able to contest old positions, and this can only happen if values are not used to protect positions from unwanted criticism. A scientist

can point to their values to argue for why they require more evidence to be convinced, but they can never point to their values to argue for why evidence is irrelevant to the claims they make or protect. Asking for more evidence drives the inquiry dialectic; holding claims above evidential critique does not.

(Douglas 2021: 30)

Perhaps a case in point would be the 1996 book *The Bell Curve* by Richard J. Herrnstein and Charles Murray which argued that African-Americans are, on average, innately less intelligent than whites, and therefore, that spending on social programs to achieve full parity and equality between Blacks and other ethnicities will be wasted. If this thesis conflicts with your social justice values, Douglas indicates that you are justified in appealing to such values in placing a very heavy burden of proof on the authors. We should not simply reject their evidence, but we can legitimately require Herrnstein and Murray to do a lot of proving. Is she right?

Can social, political, or moral values justify an epistemic demand, i.e., the demand for more evidence? Is moral outrage even an appropriate response to an empirical claim? It seems to me that it is only if you have grounds for thinking that the claim was made irresponsibly, incompetently, or in bad faith. If, for instance, there is a large body of well-founded empirical research that contradicts the Herrnstein/Murray claim, and they fail to adequately address that research, then we may indeed react with indignation. On the other hand, if we are outraged simply because an empirical claim shocks our moral sensibilities, how are we different from those proper Victorians who were just appalled that the horrid Mr. Darwin would make such imputations about their ancestry?

Douglas says that the demand for more evidence is justifiable because it "drives the inquiry dialectic." True, but only up to a point. Repeatedly raising the bar by imposing heavier and heavier burdens of proof is a common obscurantist tactic. In public debates, on climate change, for instance, moving the goalpost is a very effective delaying tactic, as Tom Toles amusingly shows in an imagined dialogue with a climate "skeptic":

Global warming is not happening

It is happening.

If it is happening, maybe it is a good thing.

It is not a good thing.

If it is not a good thing, it cannot be confirmed.

It has been confirmed.

If it has been confirmed, maybe it is not caused by humans.

It is caused by humans.

If it is caused by humans maybe it will fix itself.

It will not fix itself.

If it won't fix itself there is nothing we can do.

There are things we can do.

If there are things we can do, they are too expensive.

They are not too expensive.

If they are not too expensive, we can postpone doing them.

We can't postpone doing them.

What do you mean? This conversation has taken over a decade already.

(Mann and Toles 2016: 52)

So, Douglas's point must come with the caveat that interminable demands for "more study," or imposing an excessive burden of proof, or moving the goalposts can be just another way of refusing to consider the evidence. Asking for more evidence drives the inquiry dialectic only if we can draw a line and say enough is enough.

But, in evaluating the evidence for a hypothesis, when is there enough? That, for Douglas, is a crucial question. She notes that in evaluating hypotheses vis-à-vis evidence, we always face "inductive risk" (Douglas 2009: 58). The evidence for a scientific hypothesis, however strong, does not *guarantee* the truth of the hypothesis. The decision to accept a hypothesis therefore always carries some degree of risk. Such inductive risk leads to what Douglas calls the "inferential gap":

> It [inductive risk] points to the inferential gap that can never be filled in an inductive argument, whenever the scientific claim does not follow deductively from the evidence (which in inductive, ampliative science it almost never does). A scientist needs to decide, precisely at the point of inference . . . whether the available evidence is enough for the claim at issue. This is a gap that can never be *filled*, but only stepped across. The scientist must decide whether stepping across the gap is acceptable. The scientist can narrow the gap further with probability statements or error bars to hedge the claim, but the gap is never eliminated.
>
> (Douglas 2021: 15; emphasis in original)

Scientific claims, unlike mathematical theorems, are never *demonstrated*, but only substantiated to a degree of probability. There can never be proof, but only a preponderance of evidence. Douglas says that therefore scientists, individually and collectively, must *decide* when the evidence is sufficient. She notes that philosophers of science have appealed to "epistemic values" to close the inferential gap (Douglas 2021, 16). Epistemic values are norms that we follow because we think that by doing so we will have the best chance of achieving an accurate, objective, and undistorted understanding. For instance, we value rigorous tests of hypotheses because it is only by subjecting our hypotheses to the most severe tests that we can devise that we can have confidence in them. Douglas says that such values are useful but not sufficient:

> Note that while these virtues are very useful in assessing the strength of the available evidence, they are mute on whether the available evidence is *strong enough* to warrant acceptance by scientists. Such epistemic values do not speak to this question at all.
>
> (Douglas 2021: 16–17)

It is here, Douglas says, in serving to close the inferential gap that non-epistemic values, i.e., moral, social, or political values, play their essential role in science. However, before considering these arguments, three points need to be addressed: (1) Is closing the inferential gap a decision, a conscious act of volition? (2) Are scientists, individually or collectively, ever motivated to accept a hypothesis by epistemic (i.e., evidential) considerations alone? (3) How do we, in fact, evaluate the strength of the evidence for scientific claims?

1. Douglas speaks of scientists as "stepping across" the inferential gap and characterizes this as a psychic action, a "decision." However, it seems to me that conviction often emerges involuntarily. We are persuaded. Internal resistance to the claim suddenly crumbles and acceptance occurs in tandem when we have a moment of insight. The involuntary and

spontaneous aspect of hypothesis acceptance is what led Kuhn to speak hyperbolically of "conversion" and "gestalt shifts." I think we have probably all had experience of disbelieving something until confronted by incontrovertible evidence or the fact itself and having our disbelief suddenly and irresistibly dispelled. Belief happens.

Sometimes, indeed, we may wish that belief would not happen. In the late 1940s, scientists and others vigorously debated whether the "super" bomb, the thermonuclear bomb, should be developed (Parsons and Zaballa 2017, pp. 34–41). It was argued that a thermonuclear bomb would be a genocidal weapon, a weapon that would have no use other than sheer annihilation. Many hoped that nature would forbid such a hellish device, and the would-be developers of the "super," Edward Teller and Stanislaw Ulam, were long frustrated in their effort to contrive a workable mechanism. Alas, nature did not forbid, and Ulam's wife Françoise records the moment of realization:

> Engraved on my memory is the day when I found him [Ulam] at noon staring intense out the window in our living room with a very strange expression on his face. He said, "I have found a way to make it work." "What work?" I asked. "The super," he replied. "It's a totally different scheme and it will change history.
> (Quoted in Parsons and Zaballa 2017: 41)

Ulam's concept was improved by Teller, and quickly led to the development of the two-stage Teller/Ulam design of a thermonuclear bomb. Such weapons were possible, and if Ulam had not recognized it, someone else inevitably would have.

2. On many occasions, both in daily life and in science, the emergence of conviction plainly seems to be driven by nothing other than the evidence. In such cases there is no inference gap; the evidence suffices. It does not matter that the evidence is never strictly probative in the mathematical sense. As they say in criminal proceedings, evidence can still be "beyond a reasonable doubt." If we rid ourselves of constructivist blinders, we can see that there are many cases in the history of science where the evidence seems to have sufficed. Consider, for instance, the evidential factors that supported the acceptance of the double helix model of DNA proposed by James Watson and Francis Crick:

> Watson and Crick used the idea of the complementarity of base pairs with the two helices on the outside and the bases on the inside. [Maurice] Wilkins and [Rosalind] Franklin followed up with further confirmation from diffraction data. The double-helix structure was remarkable for the simplicity and elegance with which it revealed how genes are duplicated in heredity.
>
> (Nye 2003: 204)

Perhaps molecular biologists in the early 1950s could not specify just when such factors as the x-ray diffraction data and the virtues of simplicity and elegance would be sufficient for acceptance of the double-helix model. Yet that they *were* sufficient and that the acceptance of the double-helix model in the light of such factors was warranted, seems clear, as Ernst Mayr notes:

> Watson and Crick's double helix fitted all the facts so perfectly [The perfect fitting of the facts is probably what

"elegance" really is.] that it was accepted by everyone almost at once, even the two most actively competing laboratories, those of [Linus] Pauling and Wilkins. This dispelled all the remaining doubts whether or not DNA was truly the genetic material.

(Mayr 1982: 823)

Perhaps the acceptance of the model was not quite as instantaneous or universal as Mayr indicates, but, as biologist and historian John A. Moore notes, "During the next decade additional chemical and genetic data showed that the hypothesis [the double-helix model] was true beyond a reasonable doubt" (Moore 1993: 377). Sometimes evidence is more than convincing. Sometimes it is compelling.

3. As noted in the previous chapter in the section on Richard Bernstein's interpretation of Kuhn, the evaluation of claims by scientific communities is a process of *practical reasoning* whereby the qualified parties debate the relative merits of competing claims in the light of the evidence and in terms of our epistemic standards, canons, and values. As Douglas says, those standards, canons, and values never specify just when the evidence is sufficient. They provide no algorithms or decision procedures. However, as Marcello Pera notes in his insightful work *The Discourses of Science* (Pera 1994), scientists are often *rationally convinced* by the arguments of their colleagues, arguments that cite the evidence and apply those standards, canons, and values. The inferential gap is often closed by persuasive argument.

Maybe Douglas would concede that in some cases the evidence alone is sufficient, but she argues that in very many cases it is not, and that scientists must then make a decision, one informed by non-epistemic values, in closing the inferential gap. She states emphatically that *no* values should have a *direct* bearing on determining the claims we make (Douglas 2009: 103). Only the evidence *directly* bears on the decision to accept or reject a hypothesis. However, she says that moral and social values can have an *indirect* effect by telling us the consequences of our errors, and this information can affect our choices of methodology or guide us to closure when the evidence is ambiguous.

In the first type of case, she notes that many inquiries can result in two types of errors, false positives and false negatives. A false positive occurs when a hypothesis is accepted as true when it is false. A false negative is when a hypothesis is rejected as false when it is true. Depending upon the situation, a false negative may threaten more dangerous consequences than a false positive or vice versa. For instance, if a substance— an effective pesticide, let's say—is in fact a carcinogen at a certain concentration in the environment, then a false negative, judging that the substance is not a carcinogen when it is, could have the very serious effect of increasing the occurrence of cancer. If, however, the substance is not a carcinogen, then a false positive, judging that it is a carcinogen when it is not, could lead to unnecessary alarm as well as expensive and pointless regulation of a useful product. Further, the stringency of tests has a bearing on which type of error is more likely. Less stringent tests will decrease the probability of false negatives and increase the probability of false positives. More stringent tests will do the opposite. So, should we opt for a 95% level of significance in a given test or opt for a more stringent 99%? Douglas says that our methodological choice here should be decided by weighing all the relevant values, cognitive as well as social and ethical (Douglas 2009: 105).

Douglas admits that the choice here is difficult (Douglas 2009: 105), and, really, it is not clear that appeal to ethical and social values would be helpful unless it can be specified *whose* values are decisive. Environmentalists would say that the risk of cancer far outweighs the risk of imposing unnecessary regulations; industry lobbyists would make the opposite judgment. The general point, though, seems to be that, as is broadly recognized, scientific knowledge is not to be pursued by methodologies that contravene moral values.

To take an extreme and infamous case, the so-called Tuskegee Experiment, conducted by the United States Public Health Service and the Centers for Disease Control from 1932 to 1972, studied 400 African American men whose syphilis was left untreated to follow the effects of the disease. Clearly, such a methodology was ethically atrocious and ought not to have been conducted even if it produced genuine medical knowledge. Likewise, we might employ a less stringent methodology if a more stringent one would seriously risk animal or human suffering. Everyone will admit that scientific knowledge is valuable, but not the only value.

A more interesting case is when all the available evidence has been collected and a decision needs to be made about whether or not to accept the hypothesis:

> Was the hypothesis supported? What do the results mean? Whether to accept or reject a theory on the basis of the available evidence. Is the central choice scientists face. Often the set of evidence leaves some room for interpretation; there are competing theories and views, and thus

some uncertainty in the choice. The scientist has a real decision to make. He or she must decide if there is sufficient evidence present to support the theory or hypothesis being examined.

(Douglas 2009: 106)

In making such choices, Douglas says, scientists must consider the consequences of error (Douglas 2009: 106). If scientists decide that the evidence does support the hypothesis and then declare for the hypothesis, officials may act on that hypothesis, and the results might be good or bad. When the evidence is not conclusive, and so different evaluations are equally valid, moral values can guide us in whether to accept the hypothesis on the basis of the given evidence or to suspend judgment and await more evidence:

> Social and ethical values . . . do help with this decision. They help by considering the consequences of getting it wrong, of assessing what happens if it was a mistake to step across the inductive gap (i.e., to accept a claim), or what happens if we fail to step across the inductive gap when we should. In doing so such values help us to assess whether the gap is small enough to take the chance. If making a mistake means only minor harms, we may be ready to step across it with *some* good evidence. If making a mistake means major harms, particularly to vulnerable populations or crucial resources, we should change our standards accordingly. Social and ethical values weigh the risks and harms, and provide reasons for why the evidence may be sufficient in some cases and not in others.

(Douglas 2021: 19)

Put simply, Douglas holds that the moral and social costs of a wrongly accepting or rejecting a hypothesis can and should affect the burden of proof scientists place upon the claim. If public acceptance of a hypothesis has dire ethical or social effects if it turns out false, they should place a very heavy burden of proof on it. If, on the other hand, rejection of the hypothesis would entail such dire effects if the hypothesis is true, they should tolerate a lighter burden of proof for its acceptance. For instance, if acceptance of the Herrnstein/Murray thesis of racial inequality is likely to cause great harm if false, then place a heavy burden of proof on that claim. In the opposite instance, if failure to accept the evidence for climate change can lead to global catastrophe if climate change is real, then scientists should accept that hypothesis even if the evidence is inconclusive.

Sometimes indeed scientists need to be *very* certain before accepting a claim. Prior to the Trinity Test of the plutonium bomb in July 1945, Enrico Fermi, one of the leading Manhattan Project scientists, conjectured that a nuclear explosion would ignite the atmosphere leading to a worldwide conflagration (Rhodes 1986: 664). If you think it is possible that your experiment will blow up the world, you had better be *damn sure* before deciding that it won't!

Such a case is straightforward. Presumably, nobody wants to blow up the world. On very many other occasions, things are not so simple, and ascertaining the ethical, social, and political implications of a claim will be complex and controversial. Scientists *qua* scientists have no expertise in the weighing of often conflicting moral or social values; they are laypersons with respect to such debates. Scientists are trained to evaluate evidence; we should not expect them also to be experts in the field of applied ethics. To expect them to take the leap and make a decision on the basis of deeply disputed non-evidential considerations of which they have no special understanding is to impose an unfair burden upon them and to reduce their trustworthiness for the rest of us.

Could scientists make the decision in consultation with experts in applied ethics? However, permitting non-scientists to have a hand in saying which hypotheses scientists are to accept would be a practice ripe for all sorts of abuse, and Douglas argues that such oversight would be impossible and undesirable (Douglas 2009: 73). She claims that scientists are the only ones qualified to make the ethical decisions:

> Because science's goal is primarily to develop knowledge, scientists invariably find themselves in uncharted territory. While the science is being done, presumably only the scientist can *fully* appreciate the potential implications of the work and, equally important, the potential errors and uncertainties of the work. And it is precisely these potential sources of error, and the consequences that could result from them, that someone must think about. The scientists are usually the most qualified to do so.
>
> (Douglas 2009: 73–74; emphasis in original)

Yet, it is one kind of judgment to assess the uncertainties of scientific work and an entirely different kind of judgment to assess complex moral and social consequences of those uncertainties, and, as noted, scientists are experts in the former kind of assessment but not the latter. Sometimes weighing the moral and social consequences of scientific claims may be straightforward and uncontroversial, but very often such adjudications will be complex, nuanced, and multi-sided—not the kinds of decisions that should be left to amateurs. And, once again, *whose* values get to decide for scientific communities? A scientist working for Greenpeace will have one set of values, and one working for Monsanto will have another. So, requiring scientists to consider non-epistemic values, instead of facilitating decision-making, seems likely to often bring further indecision. If closure is often difficult to reach on evidential grounds, so, *a fortiori* will be closure on values considerations.

Douglas suggests that with respect to the highly controverted and complex issues, scientists could maintain a position of value neutrality (Douglas 2009: 124). Two problems: First, scientists are probably no more capable of neutrality on controversial issues than the rest of us. Second, if such a neutrality were achievable, what guidance could it give to the kinds of choices Douglas thinks values should impact? If one side wants more evidence and the other side thinks we have plenty, what do you do from a neutral position? Perhaps, though, neutrality is not an individual achievement, but a collective one. Douglas indicates this possibility by considering the consensus attained by the debates among members of the Intergovernmental panel on Climate Change. These conferences involved thousands of scientists from 150 different countries with a wide range of social and ethical perspectives (Douglas 2009: 131). However, neutrality achieved among such a very diverse group would seem to involve a *washing out* of the impact of social and political values as multifarious perspectives correct each other by pointing out *objective* considerations that value commitments caused others to overlook. In fact, this process seems to be the whole rationale for calling for a diversity of inquirers, yet it is the opposite of a process of letting ethical or social values carry us across the inferential gap.

Most basically, what is a scientific hypothesis and what does it mean to accept one? A hypothesis is our proposal that something *is so*. When we accept a hypothesis we accept that its claim is true, or, at least, that the claim is the most rational, justified, or warranted of all the competing claims we have. If our decision to endorse a hypothesis is based—directly, indirectly, or to any degree whatsoever—on factors that have no bearing on the truth, rationality, justification, or warrant (i.e., the *epistemic* status) of the hypothesis, then, unavoidably, that endorsement is no longer only about *what is*, but, in part, is about *what we prefer*. Maybe this is no problem when it is something that we all prefer, like not blowing up the earth. It is a problem when the preferences are highly disputed and passionately debated.

The reason that books like this one need to be written is that the trustworthiness of science has been under vociferous attack and the authority of scientists radically undermined. If scientists *qua* scientists (and not as public intellectuals) become embroiled in debates about hot-button issues, their authority will erode even more. Scientists who demand a higher burden of proof because of their values will be accused of intransigence by those with other values. Those whose values prompt them to accept a lighter burden will be accused of surrendering to an ideological motivation to jump to conclusions. In fact, it is precisely the purported pollution of science by wrong values that critics of science such as Harding on the left and Johnson on the right charge in their attacks. Relegating values to an "indirect" role will not silence such critics.

What, though, if our ethical values sometimes *do* possess epistemic credentials? This interesting suggestion is made by Matthew J. Brown, one of the commentators on Douglas's Descartes Lectures (Brown 2021: 51–65). He draws on distinctions made by John Dewey:

... following the work of John Dewey is the distinction between *valuing* and *evaluation, prizing* and *appraisal*, between what we happen to like or dislike and the values certified by a process of *value judgment*. Our pre-reflective values may lack evidential support, as do many of our habits, biases, assumptions, and received beliefs. Our value judgments, in the honorific sense of "judgment," are the product of reflection and inquiry, and as such (according to Dewey), value judgments must have evidential support. Our well-established value judgments have been tested in practice under a variety of situations, and led to success—social, practical, emotional, and scientific.

(Brown 2021: 53)

Consider feminist values. Have they guided science in the right direction? Brown says that they have a good track record:

Feminist values have a good track record of guiding science into productive channels, whereas sexist values have the opposite sort of record. This fact lends feminist values further support. What's more, the introduction of feminist structures in many fields has had a positive transformative effect on those fields.

(Brown 2021: 54)

In the same volume, Kristina Rolin notes instances in which feminist values prompted positive changes in science:

... when feminist scientists entered the field of human evolution, they introduced a novel perspective on the anatomical and behavioral development of the human species. In the controversy over human evolution in the

1970s, they challenged the "man the hunter" narrative by developing "the woman the gatherer" narrative to offer an alternative interpretation of empirical evidence.

(Rolin 2021: 43)

Douglas comments on how feminist values have corrected scientific errors:

> Feminist philosophers of science showed how sexist values blinded science to alternative explanations of phenomena or directed the attention of scientists to some narrow subset of data, a fuller explanation of which produced rather different interpretations and results. Examples from archaeology (explanations of how tool use developed), cellular biology (explanations of fertilization processes), and animal biology (explanations of duck genital morphology and mating behavior) demonstrate such influences of values on behavior in spades.

(Douglas 2021: 13)

Duck genitalia aside, I think that a fair examination reveals that the influence of feminist values on science and other fields of knowledge has been a very mixed bag. Paul Gross and Norman Levitt have fun debunking some of the sillier and wilder claims made in the name of feminism, such as the promotion of a "feminist algebra" and the metaphor mongering that purports to be an exposé of sexist bias in biology (Gross and Levitt 1994: 107–148). But isn't this unfair? Instead of picking off outliers, shouldn't a serious critique focus on the claims of "mainstream" feminism? However, as Gross and

Levitt's critique shows, it is not always easy to distinguish the mainstream from the fringe.

Consider the case of Marija Gimbutas. Gimbutas was an academically trained archaeologist and anthropologist who held distinguished academic appointments at Harvard and UCLA. Gimbutas is best known for her claim that prehistoric Europeans lived in peaceful, egalitarian societies that worshipped the Mother Goddess and were ruled by wise matriarchs. This Eden was destroyed when men took over in the Iron Age and imposed patriarchy with its concomitant hierarchies, warfare, and subjugation of women (Gimbutas 1991). Gimbutas' followers have popularized in novels and movies the myth of the Mother Goddess and her pre-patriarchal paradise.

Speaking personally, I am sympathetic to the revival of ancient religiosity, and have recently written in defense of neopaganism (Parsons 2022). However, as an archaeological thesis, Gimbutas' claim has been widely and deeply criticized, as classicist Bruce Thornton notes (Thornton 1999: 196–207). In fact, It does not seem overly harsh to characterize Gimbutas' scenario as "religious myth disguised as scientific fact" (Thornton 1999: 213). The theme of a fall from a Golden Age is an ancient and powerful one, and it is unsurprising that feminists would want their own version of that myth. Yet it is hard to see how Gimbutas and her followers are really doing anything very different from what fundamentalist Christians do in taking the Genesis narratives as literal history. In both cases myth is turned into pseudoscience and only obscurantism can result.

Naturally, it is in the study of sex and gender that feminist influence is strongest. Steven Pinker identifies "gender feminism" as a dominant mode of feminism in his book *The Blank Slate* written in 2002 (Pinker 2002: 341–343). He says that gender feminism is committed to three empirical claims:

> The first is that the differences between men and women have nothing to do with biology but are socially constructed in their entirety. The second is that humans possess a single social motive—power—and that social life can be understood only in terms of how it is exercised. The third is that human interactions arise not from the motives of people dealing with each other as individuals, but from the motives of *groups* dealing with other groups—in this case the male gender dominating the female gender.
>
> (Pinker 2002: 341; emphasis in original)

Pinker says that these claims are contrary to findings in neuroscience, genetics, ethnography, and psychology (Pinker 2002: 341). If Pinker is right, at least one major mode of feminism has placed itself directly in the path of scientific progress.

Have things changed for the better in the 20 years since Pinker's book? Fast forward to 2020 and Debra Soh's *The End of Gender* (Soh 2020). Soh is a journalist and also an academically trained, Ph.D. neuroscientist and sexologist who has published in academic journals. She comments:

> Mainstream feminism has been very effective in spreading its tenets about gender and is now sowing [I think she means "reaping"] the fruits of its labor. In the name of affording girls and women the same opportunities and rights as boys and men, news stories, educational institutions, and governmental policies have taken to

> broadcasting a similar message: men and women are, at the core, the same, and any differences we do see are due to socialization and sexism.
>
> (Soh 2020: 40)

She counters:

> Scientific studies have confirmed sex differences in the brain that lead to differences in our interests and behavior. These differences are not due to the postnatal environment or social engineering. Gender is indeed biological and *not* due to socialization.
>
> (Soh 2020: 41; emphasis in original)

There are quite a few such critics, many with seemingly impeccable feminist credentials.[2] Can they all be dismissed as incompetent researchers, tools of the patriarchy, or faux feminists? If not, then their arguments have to be taken seriously, and if they are even partially correct, we may conclude that feminism has not always guided science down the best paths. This would be an utterly unsurprising conclusion. An ideology is a creed, a preformed set of beliefs taken as definitive and authoritative. In that sense, feminism is as much an ideology as religious fundamentalism, Marxism, or libertarianism. Creeds may intersect reality at some points and diverge at others, so it is expected that creedal claims would sometimes guide truly and at other times not. This is not to say, of course, that feminism is monolithic; there are many feminisms. Hence, blanket criticisms—or blanket endorsements—are not appropriate. However, it appears that many prominent forms of feminism—even "mainstream" ones—have played obscurantist roles.

In general, I agree with Brown's claim that value judgments can have epistemic credentials, and that those credentials can, in principle, make them relevant to scientific practice. However, it is not clear to me that feminist claims consistently have such epistemic credentials, and some appear to be plainly dogmatic and obscurantist. Therefore, the recommendation, as a general policy, that science should be guided by feminist values seems unwise. This conclusion should reinforce our caution about letting our favored causes get too cozy with our science.

I have only touched on a very few aspects of Douglas's very rich study of values and science. There is much that she says with which I can agree. However, the idea that scientists should intentionally, even if "indirectly," incorporate moral and social values into the assessment of factual claims, strikes me as true in principle but highly problematic in practice. When the social, moral, and political issues surrounding a scientific claim are highly complex and controversial, scientists do their duty when they speak where the evidence speaks and stay silent where the evidence is silent. The responsibility then falls upon the rest of us to be much more informed and critical consumers of scientific information. We crave certainty, especially on issues, like climate change, that affect all of us. However, science often leaves us with a considerable degree of uncertainty. Fortunately, the field of decision theory provides us with rational ways of making decisions under uncertainty. *There* is the context where values enter in crucially and necessarily.

Finally, I am not in any way advocating that science be "value free." Douglas argues that there is no clear line of demarcation between epistemic values and moral values (Douglas 2009: 89–91). I will go one better and say that there is *no* distinction between epistemic values and moral values because epistemic values ARE moral values, though not all moral values are epistemic values (i.e., epistemic values are a subset of moral values). As rational beings we have a moral *duty* to be rational, that is, to do our best to proportion our beliefs to the evidence and to be willing to modify or reject our beliefs when the evidence is against them. This duty applies with special rigor to scientists.

The only reason that science has any authority at all is that it is *here*, preeminently, that evidence has to count more than anything. A scientific claim is respected precisely because we trust that—for once—the clamors of dogma, politics, and special interests are muzzled, and that every effort has been made to let *nature* have the final word. Any suspicion that a scientific claim is based not solely on scientists' best understanding of *what is*, but to any degree on what they *wish to be*, inevitably undercuts the authority of science. When, therefore, a scientist decides to endorse a hypothesis, but there is legitimate suspicion that the decision was to any extent motivated by considerations—however laudable in themselves—not pertaining to the truth, warrant, or promise of that hypothesis, then there will inevitably be an appearance of dereliction of scientific duty. Science is too precious and fragile a good to be weakened by such suspicions.

Dinosaur Revolutions

6

Before he became a historian and a philosopher of science, Thomas Kuhn was trained as a physicist, taking a Ph.D. from Harvard. Naturally, then, many of his historical examples are drawn from the history of physics and astronomy, and most of the rest from chemistry. The history of these fields is interesting and important, but I like dinosaurs. From a very early age, dinosaurs engaged my imagination, and I avidly collected dinosaur books and toys (I still do). I am still the proud possessor of *The Giant Golden Book of Dinosaurs* published in 1960, with magnificent illustrations by Rudolph F. Zallinger, painter of the great mural *The Age of Reptiles* at Yale's Peabody Museum. I remember my excitement at age seven getting to see *Journey to the Center of the Earth* with actors battling a Dimetrodon (not a dinosaur, but close enough). Unlike most juvenile dinosaur enthusiasts, my fascination has continued into adulthood.

In my lifetime—since the publication of my childhood dinosaur books—the understanding of dinosaurs has changed radically. Some say that it is a "revolution" in dinosaur paleontology. Others go even further and claim that the emergence of new methods, and, in fact, whole new fields, has transformed paleontology into a genuine science. However characterized, there is no question that many claims about dinosaurs now widely (but not universally) accepted among paleontologists

DOI: 10.4324/9781003105817-7

would have seemed astonishing, heretical, or preposterous 60 years ago. Among these claims are the following:

1. Birds are evolutionary modified theropod dinosaurs. Conversely, the non-avian dinosaurs, those "terrible lizards" were much more birdlike than lizard-like. Many had feathers and may have had a high-energy metabolism closer to that of birds than reptiles.
2. *Tyrannosaurus rex* lived fast and died young, rapidly growing from a small hatchling to an eight-ton adult and dying before age thirty. *T. rex* had a big brain and highly developed senses. It may have been smarter than dogs or cats and may have hunted in packs.
3. The dinosaurs did not die off gradually due to the slow deterioration of their environments and competition with allegedly superior mammals. Rather, the extinction of the dinosaurs was genuinely sudden and caused by the catastrophic impact of a large asteroid or comet at the end of the Cretaceous.

Dinosaur paleontology thus presents a remarkable set of case studies of how new scientific ideas emerge and how they are argued out. It is therefore an excellent locus to consider some of the controversial claims about science considered and examined in the first four chapters. In particular, in this chapter I will seek to answer the following three questions:

1. Are dinosaurs social constructs? That is, are we justified in thinking that we really know some things about dinosaurs, or are our ideas figments of the paleontologist's imagination, constructs shaped by politics and ideology? What is the nature of the methods and that brought about this new

understanding of dinosaurs? Can we trust that those methods reliably indicate what dinosaurs were really like, or are they products of social conventions?

2. How did we learn that birds are dinosaurs? We now know that birds did not just evolve from dinosaurs, but that they are the dinosaurs that survived the end-Cretaceous event that destroyed the non-avian dinosaurs. How did we learn this startling fact?

3. The shift from an earthly and gradual cause of the end-Cretaceous extinction to an extraterrestrial and sudden one challenged assumptions that had reigned for 150 years. Here, surely, was something very much like a Kuhnian paradigm shift, with a radical new claim that brought not only new evidence, but new standards and values. So, did the opposing sides experience an *in-principle* breakdown in communication, so that their discourse became incommensurable? Did those, like eminent paleontologist David Raup, who came to accept the new catastrophist paradigm, do so by a gestalt-switch like "conversion," or were they simply persuaded by what they took to be solid evidence and the application of rigorous methods?

THE RISE OF PALEOBIOLOGY: MORE THAN STAMP COLLECTING

In 2009, an important book was published by the University of Chicago Press titled *The Paleobiological Revolution: Essays on the Growth of Modern Paleontology*, edited by David Sepkoski and Michael Ruse (Sepkoski and Ruse 2009). The book contains 26 essays on the rise of modern paleobiology, the science of the biology of extinct organisms. Whether this development should be called a "revolution" in Kuhn's sense or not is not clear. Kuhn might have considered it the emergence of a paradigm from a pre-paradigm state rather than the replacement of one paradigm by another. In any case, the effects of these new studies have been profound, replacing ignorance and speculation about the lives of extinct creatures with testable hypotheses and multiple new lines of evidence.

Circa 1920, Nobel laureate Ernest Rutherford (1871–1937) famously sneered that any science other than physics was mere "stamp collecting" (Benton 2019: 11). He meant that only physics produced deep theories stringently testable by hard data. The other so-called sciences are mostly collected and categorized without providing deep insights into the nature of things. However unfair such a characterization might have been of chemistry or biology, it was painfully close to home with paleontology. Paleontologists were very good at collecting fossils, reconstructing the skeletons of extinct creatures, and naming them, but with respect to most questions about dinosaurs as living creatures, they could do little more than guess and speculate. The attitude of evolutionary biologists towards paleontology was one of condescension, with little expectation that paleontologists could have much to say of importance about evolution (Sepkoski and Ruse 2009: 1–5).

Self-described "dinosaur heretic" Robert Bakker, in his many television appearances, has denounced the orthodox view of dinosaurs as "slow, stupid, and in the swamp." He counters that we now know that dinosaurs were "fast, smart, and on the land." There were always dissidents who viewed the dinosaurs as lively and energetic creatures, but, overall, Bakker seems to have been right. Paleontologists, with little real data to go on, assumed that gigantic reptilian creatures must have been torpid and stupid. Artistic reconstructions of dinosaurs clearly reinforced such an image. The huge sauropods such as *Brontosaurus* and *Brachiosaurus* were nearly always depicted as neck-deep in water, as though their enormous bulk could only be supported by being buoyed up. The whole image was one of hulking inadequacy, thereby supporting the metaphor, still used today, of a "dinosaur" as anything obsolescent, outmoded, and fit only for replacement.

Developments in paleobiology from the 1970s on have produced a new image of dinosaurs as living animals. Because space is limited, I will not offer an overview but will focus on that rock star of the Cretaceous, *Tyrannosaurus rex*, undoubtedly the most famous dinosaur. *T. rex* and its relatives such as *Albertosaurus, Gorgosaurus, Daspletosaurus,* and *Tarbosaurus* lived in the last several million years of the Cretaceous and were the apex predators in their habitats. From the first discovery of its fossils by Barnum Brown in 1902, and its scientific description by Henry Fairfield Osborn of the American Museum of Natural History in 1905, it was clear that *T. rex* was a fearsome creature. It was huge—now estimated at over 12.3 meters (40 feet) long and weighing 7.7 metric tons (nearly 17,000 pounds). Its skull was the size of a bathtub with a mouth full of banana-sized teeth, serrated like steak knives (Benton 2019: 237).

In my childhood dinosaur books, *T. rex* was depicted as a sluggish creature that mostly rested its enormous bulk and would only stir when driven by hunger to seek a meal. After a gigantic gorge, it would return to a state of torpor. In fact, 60 years ago there was little information about dinosaurs as living creatures. Paleontologists could only speculate about their physiology, growth, intelligence, behavior, and lifestyle. Now they have considerable evidence relevant to a number of those details. What evidence do we now have about how *T. rex* and its slightly less enormous relatives lived?

Let's start with T. rex's brain. Famously, some dinosaurs were tiny-brained. In dinosaur lore, multi-ton *Stegosaurus* has always been derided as having a brain the size of a walnut. Brains are metabolically expensive items, so having more brains than is needed for an organism's lifestyle would be a wasteful luxury. T. rex had a big brain that clearly suited its needs. How do we know? Soft tissues like brains are not preserved in the fossil record. CAT scans, which are high-powered X-rays, can be used to examine the brain cavity in dinosaur skulls.

One thing we can learn from the scans is the encephalization quotient (EQ) of T. rex (Brusatte 2018: 219). Put simply, EQ is a measure of the relative sizes of brains of different organisms adjusted for differences in body size (Brusatte 2018:219; for a technical definition of EQ see Hurlburt, Ridgely, and Witmer 2013: 137). After all, brain size and body size are closely correlated so that big animals will tend to have larger brains just because they are bigger, and not because they are smarter. EQ corrects for that correlation. A small animal can have a smaller brain than a bigger animal in terms of absolute size, but its brain can be larger in proportion to its body size.

EQ is used as a proxy for making rough comparisons of the intelligence of different animals. The higher the EQ, presumably the more intelligent the creature is. The EQ for humans is 7.5, for dolphins 4.0 to 4.5, for chimps 2.2 to 2.5, and for dogs and cats 1.0 to 1.2. T. rex had an EQ of 2.0 to 2.4 (Brusatte 2018: 219), so the tiny-brain stereotype did not apply to T. rex. These results support the hypothesis that T. rex had considerable cognitive capacities and was capable of a range of complex behaviors.

Perhaps more revealing than EQ estimates are the clues that T. rex brains give about its sensory abilities. T. rex had enormous olfactory bulbs, lobes so large that, even when their size is normalized for T. rex's great body mass, that, according to Brusatte, they are extreme outliers when compared to other predatory dinosaurs (Brusatte 2018: 219–220). The Tyrant King clearly had a very acute sense of smell. Other senses were well-developed also. The CAT scans show that T. rex's inner ear had an elongated cochlea, and such a structure in living animals is an adaptation for sensitivity to low-frequency sounds (Brusatte 2018: 220). The inner ear also controls balance, and its structure indicates that T. rex was capable of highly coordinated movements of the head and eyes and, despite its bulk, was capable of agile movement (Brusatte 2018: 220). Finally, T. rex had enormous eyeballs, positioned so that they were capable of binocular vision and depth perception (Brusatte 2018: 220). Dinosaur paleontologist Steve Brusatte sums up: "Thus it wasn't all brute strength. T. rex had brawn all right, but it also had brains. High intelligence, world-class sense of smell, keen hearing and vision" (Brusatte 2018: 220).

T. rex therefore had the acute senses a predator needs to locate and track prey and the brute strength to subdue it. Like the great white shark, T. rex killed with its enormous bite. How did those terrible jaws work, and what can be inferred from that information about T. rex's behavior? Just how powerful was that bite?

Paleontologist Paul Molnar did an exhaustive study of the mechanics of the tyrannosaur bite and the forces it could exert. Jaws are basically levers, with muscles applying twisting force—torque—to the mandible, the lower jaw, to give it rotational acceleration around a fulcrum—the jaw joint (the craniomandibular joint). Torque is determined by the product of force applied to the lever and the perpendicular distance from the fulcrum to the line of action of the applied force (called the "lever arm"). The concept of a lever arm is easily illustrated. If you want to loosen a tight nut with a wrench, and you push straight down on the wrench, the lever arm is the distance between the nut and the point on the wrench where you are applying the force. You get more leverage the longer the lever arm, that is, the farther from the nut you grip the wrench.

Since the forces exerted by the muscles of extinct animals cannot be directly measured, Molnar estimated the contribution of each set of jaw muscles to the overall torque of the bite by determining the lever arm of the jaw muscles (Molnar 2013: 179). This permitted him to graph the gape of the jaws of various theropods versus the lever arm of different muscles to see how the lever arm of the various jaw muscles changed with respect to the degree of gape (Molnar 2013: 181). Molnar concludes that the tyrannosaurs had a more powerful bite than other theropods such as *Ceratosaurus* and *Allosaurus*, and that the greater bite efficiency of tyrannosaurs may have played a role in their replacement of the allosaurids as apex predators in North America (Molnar 2013: 191). Further, because of its larger size, and consequently bigger muscles, *Tyrannosaurus rex* would probably have had a significantly stronger bite than the other tyrannosaurs (Molnar 2013: 190).

Molnar's study of bite dynamics gives information about the relative strength of the tyrannosaur bite, but Brusatte notes more direct evidence. He reports on experiments done by Steve Erickson of Florida State University, who made a bronze and aluminum cast of a *T. rex* tooth, put it in a hydraulic loading machine, and had it clamp down onto the pelvis of a cow (Brusatte 2018: 205–206). The bone of a cow pelvis is very similar to that of a *Triceratops*, and the aim was to see how much force would be required to produce a puncture in the cow bone as deep as the puncture a *T. rex* tooth had made in a

Triceratops bone. The force required was 13,400 newtons, about 3000 pounds of force exerted by just one tooth (Brusatte 2018: 206). Other evidence indicates that this was far from the maximum bite force *T. rex* could exert.

Michael J. Benton reports studies done by Emily Rayfield of Bristol University that employed the engineering method of finite element analysis (FEA). Engineers and architects use FEA to stress-test models of structures. Rayfield employed it to determine the distribution of forces in the *T. rex* skull when it exerted its enormous bite (Benton 2019: 190). Using a digital model of the *T. rex* skull, she estimated that each tooth could exert a maximum pressure of 31,000 newtons (Benton 2019: 192). Other researchers using other methods got similar results:

> In a further study using a different biomechanical computing approach called multi-body dynamic modeling, Karl Bates and Peter Falkingham estimated a range of bite force values from 35,000 to 57,000 newtons, or the equivalent of 3.6 to 5.8 tonnes [metric tons] of weight. This is the strongest bite force ever demonstrated in any animal living or extinct, far greater than that of the living Great White shark . . . Importantly, all these approaches give similar values which suggests that there is a good chance they might be correct.
>
> (Benton 2019: 192)

T. rex was well-adapted to the role of a dinosaur cruncher, with acute senses and overwhelming power. Just how, though, did it hunt? Some have argued that, in fact, *T. rex* was a scavenger and not an active hunter (see Horner and Lessem 1993). So, was it a hunter, and if so, how did it stalk its prey?

So, was T. rex a scavenger or an active hunter? Actually, this way of posing the question is simplistic. All carnivores will scavenge given the opportunity. Why pass up a free meal? The question is whether T. rex was an obligate scavenger, like the vulture. Did it have to scavenge, being incapable of active hunting? Arguments for T. rex as an obligate scavenger include the following: (1) T. rex was not a fast runner and could not have chased down its prey. (2) Tyrannosaurs have tiny arms in proportion to their size, and it is hard to see how they could have been used in predation. (3) T. rex's acute sense of smell, as indicated by its oversize olfactory bulbs, would have been very useful for detecting carrion. (4) T. rex appears to have had poor eyesight, and this would have been a disadvantage for an active hunter, but not a scavenger.

Kenneth Carpenter addresses these claims. How fast could T. rex run? One way of estimating an animal's running abilities is to compare the relative lengths of the femur, the upper leg bone, to the tibia, the larger of the two lower leg bones. Animals built for speed have a lengthy tibia with respect to the femur; slower animals have a lower tibia-to-femur ratio. A horse, for instance, has a tibia as long as its femur. Humans, on the other hand, have tibias only 73% of their femur length. Even Usain Bolt is not built for speed the way the horse is! T. rex had a tibia 92% of the length of its femur. Further, being digitigrade (toe-walking), the metatarsals (the long bones in the foot) also need to be considered, and in T. rex they are 50% of the length of the femur (Carpenter 2013: 269). Further, elastic energy stored in the tendons of the feet and tail would have been released as the animal ran, pushing it forward (Carpenter 2013: 270–271). Compared to T. rex, Edmontosaurus and Triceratops, potential prey species, had significantly lower tibia- to-femur ratios (Carpenter 2013: 269). T. rex did not have to be especially fast, only faster than its prey. Further, healed injuries evident in the skeleton of a hadrosaur indicate that it survived an attack while fleeing a large theropod, possibly the tyrannosaurid *Daspletosaurus* (Murphy, Carpenter, and Trexler 2013: 279). So, there is fossil evidence of predator/prey chases.

Concerning the infamous *Tyrannosaurus* forelimb, it was short relative to the animal's great body size, but it was actually as long as a human arm though much stronger with far more massive and heavily muscled bones (Carpenter 2013: 272). Further, forelimb injuries and pathologies are of just the sort that would be produced by the effort to control violently struggling prey (Carpenter 2013: 272).

As for *T. rex*'s senses, it has already been mentioned above that *T. rex* had large eyes, and Carpenter adds additional evidence (Carpenter 2013: 268). As noted, Brusatte claims that *T. rex* was capable of binocular vision and depth perception (Brusatte 2018: 220). As for *T. rex*'s olfactory abilities, Carpenter notes that crocodilians also have large olfactory lobes, as do some birds that are active hunters (Carpenter 2013: 267). Further, dinosaurs in general tended to have large olfactory lobes (Carpenter 2013: 268). Finally, a keen sense of smell could be as useful to an active hunter as to a scavenger (Carpenter 2013: 268).

It appears that *T. rex* was well adapted to be an active hunter, but what hunting technique might it have used? David A. Krauss and John M. Robinson provide a biomechanical analysis that shows how *T. rex* could have employed a hunting strategy to subdue even such fearsome prey as *Triceratops*. They amusingly compare it to the bucolic pastime of "cow tipping," that is, they show how *T. rex* could have attacked from the side with enough force to knock even a large *Triceratops* off its feet (Krauss and Robinson 2013: 251). They comment:

> The most interesting aspect of this theory is that it explains most of the unique features of tyrannosaur anatomy.

Specifically, their small arms seem to be an adaptation allowing them to grasp their prey's back while pushing it with the pectoral region of their torso. Their large heads would have helped in tipping their prey over and their large mouths and bone-piercing teeth would have made bites to the side more effective and lethal. According to this "ceratopsian-tipping" hypothesis, *Tyrannosaurus* would have ambushed its prey from cover, knocking it over and rendering it vulnerable, then killing it with a swift bite to the rib cage.

(Krauss and Robinson 2013: 251)

We have then a variety of lines of evidence that indicate that T. rex was an active predator, capable of subduing even the large and powerful prey species of its environment. Instead of a sluggish and stupid creature, we have the image of an agile, canny predator that ambushed its prey from cover and vanquished it with a gigantic bite. In short, T. rex was every bit as fearsome as imagination has conceived it: a living nightmare, like a great white shark on land, only much bigger and more powerful.

So, how certain can we now be that we know what T. rex was really like? Are the noted methods reliable indicators of T. rex's true nature? When we talk about fossil creatures, imagination must always supplement fact. As one paleontologist put it to me, we are far from knowing everything about living creatures, so we surely cannot know all about extinct ones. Further, as David E. Fastovsky notes, our interpretations of dinosaurs are inevitably conditioned by social context (Fastovsky 2009: 239). As evidence, Fastovsky cites early interpretations of T. rex that emphasize power and domination, reflecting the imperialism and colonialism of American and European powers of the day (Fastovsky 2009: 240–241). By contrast, Fastovsky notes, G. S. Paul, writing in the very different *Zeitgeist* of the 1980s, depicts *T. rex*, not as the solitary, stereotypically male hunter of the early depictions, but as a social creature that hunts cooperatively and exhibits a maternal concern for child rearing (Fastovsky 2009: 244). Fastovsky does not say that this new view of *T. rex* is a pure invention. There are new discoveries, but those discoveries have occurred in a social milieu that was receptive to them:

> It is true that with new discoveries come new insights, and certainly new discoveries contributed to modern interpretations of tyrannosaur behavior. These interpretations are indeed *modern*, but not necessarily *timeless*. Had the cultural ground not been fertile, the significance of these discoveries and insights might not have been appreciated. Why were the first assessments of *T. rex* that of a solitary, dominant predator? Why are the latter assessments so different? The change resulted partly from a paradigm shift from the crocodile model of dinosaur metabolism to something more birdlike. But it is no coincidence that the biggest advances in our understanding of *Tyrannosaurus* parallel the cultural milieu in which those advances were made.
>
> (Fastovsky 2009: 245; emphasis in original)

It is very easy for interpretations of dinosaurs in terms of dominant cultural motifs to get overblown, as indeed they do, for instance, in W.J.T. Mitchell's *The Last Dinosaur Book* (Mitchell 1998). For Mitchell, dinosaurs are iconic projections of various cultural, psychological, and political obsessions. Mitchell's and similar analyses, which appear to rely on methods no deeper than free association, so conflate fact and fiction that they lose track even of the basic fact that dinosaurs were *animals* (for a detailed examination of Mitchell, see Parsons 2001: 106–125).

Fastovsky, of course, does not go that far, but he does suggest (Fastovsky 2009: 252) that methodological innovations may be the product of social and political agendas. If he means only that the value and significance of good methods are more readily appreciated in certain social milieus than others, then this seems plausible. If, on the other hand, he is suggesting, similarly to Shapin and Schaffer in *Leviathan and the Air Pump,* that the paleobiological methods are adopted, not because they are good methods, but merely to satisfy social or political agendas, then this would be very implausible.

For instance, should we perhaps think that the methods employed by Emily Rayfield such as FEA and computer modeling are respected because, say, traditionally male-dominated paleontology has developed a feminist conscience, and so is especially receptive of women's contributions? Here is what Rayfield says about methodological developments in studies of how dinosaurs fed:

> When I started my research we had some information from the fossils, such as tooth shape, tooth marks, stomach stones, and coprolites. A few experts in biomechanics had suggested ways to model dinosaur jaws like levers, so you could make some basic calculations, but we now have integrated computational methods that allow much more complex—or realistic—questions to be asked.
> (quoted in Benton 2019: 213)

I suspect that Rayfield would regard it as farfetched—not to mention patronizing—to suggest that the acceptability of her methodological innovations was due chiefly to her femaleness rather than the fact that they are better methods.

Why think that Rayfield's and the other methods considered are good methods, that is, that they probably give us information about what dinosaurs were really like? Rayfield employs FEA and computer modeling techniques that have proven their effectiveness in other fields. Molnar's studies of the tyrannosaurid bite were based on the simple physics of torque and levers applied to detailed anatomical knowledge. Brusatte used CAT scans to precisely image T. rex's brain and based his conclusions about its sensory abilities on well-known facts of neuroanatomy and analogies with living animals. Erickson experimented with a model T. rex tooth and measured the biting force required to puncture a cow pelvis as deeply as T. rex had bitten into Triceratops bone. Krauss and Robinson did a biomechanical analysis drawing upon basic physics to provide a plausible hunting strategy for T. rex. Carpenter provides an inference to the best explanation to explain the observed injuries and pathologies of the T. rex forelimb.

Prima facie, then, these methods offer promise of more and better data and the more rigorous evaluation of hypotheses about T. rex. These diverse studies are mutually reinforcing, and so our picture of T. rex is now both more empirically grounded and epistemologically more coherent. Are these methods conclusive? That is not for a layperson to say. Paleontologists must argue out the value of these methods as they do with all methods. There are always assumptions and inferences that can be challenged, data that can be questioned, and alternate accounts that need to be considered, and only those with the requisite expertise can do this.

And that is precisely the point. To argue that a proposed method is not a good method, it is insufficient to tag it as politically appealing to those who accept it. Unless you dismiss *a priori* the legitimacy of all scientific methods, as Shapin and Schaffer do, you have to get down into the trenches and show that the method does not deliver the epistemological goods. In other words, you must enter the scientific debate. However, in so doing you are tacitly admitting that there is more involved in the evaluation of scientific methods than ascertaining their political efficacy.

In fact, if it is a question of providing warrant for its claims, it does not matter at all whether a certain scientific method bears the imprimatur of a particular political preference or social agenda. All that matters is whether it is a good method, i.e., whether it provides the evidence it should. If the method *really does* deliver the goods, then, with respect to warrant, its political expedience is irrelevant.

How, though, can we be confident that scientists accept methods *because* of their cognitive worth rather than their political value? No matter how politically motivated you are, you want to support your views with good arguments and good evidence—the best you can get. Scientists are very good at critical thinking, and they see right through rhetorical bluster and are intolerant of hand waving. Scientists' culture of skepticism subjects even politically appealing hypotheses to rigorous critique. There are no free passes for hypotheses by women or minorities; if anything, they probably still have to work harder to be taken seriously. Besides, scientists are anything but monolithic in their political views. I have known scientists that were far right, far left, and everything in between. If you want to convince an ideologically diverse community of colleagues, you had better have your logic tight and your evidential ducks in a row.

In short, we do seem to know something about T. rex. That knowledge may not be "timeless" as Fastovsky put it, but, of course, no scientific hypothesis is. All hypotheses are, in principle, revisable in the face of new evidence. That is a fact that makes science more trustworthy, not less.

THE DINOSAURS TAKE WING

Dinosaurs with feathers? Not long ago the notion would have seemed ridiculous. Dinosaurs were reptiles, and reptiles do not have feathers. Only birds have feathers.

Then they found the feathered dinosaurs. In recent decades, many of the most important dinosaur fossils have come from China. In 1996 in the northeastern province of Liaoning, a small dinosaur was found and given the name *Sinosauropteryx*. It was quite similar to the familiar small theropod *Compsognathus*, but it seemed to have hairlike fibers projecting from its neck, spine, and tail. Were these protofeathers? Opponents of the dinosaur-to-bird hypothesis dismissed these hairy fibers as muscle tissue, or perhaps a reptilian frill or sail. The next year, though, a much better specimen was found, and the fibers were found over the whole body, so they could not have been a frill or sail.

In June 1998, the leading science journal *Nature* carried an article by Ji Qiang, Philip Currie, Mark Norrell, and Ji Shu-An reporting the discovery of a dinosaur with unmistakable feathers. This creature was named *Caudipteryx*, and it had feathers that were not hairy fibers but like those of birds with a central shaft and parallel barbs projecting from opposite sides. Some paleontologists dismissed *Caudipteryx* as a bird, but later discoveries, such as *Microraptor gui*, a small theropod with flight feathers on both hind and forelimbs. (For a fuller account of these discoveries see Benton 2019: 118–124.)

Actually, the similarities between birds and small theropods had been noted by Richard Owen. Owen, the staunch antievolutionist, must have been mortified when his brilliant and bitter antagonist Thomas Henry Huxley then used that evidence to argue for the evolution of birds from dinosaurs (Benton 2019: 114). The discovery of *Archaeopteryx lithographica* just a year after the publication of Darwin's *Origin of Species* seemed to provide the perfect "missing link" between dinosaurs and birds. In fact, *Archaeopteryx* is such powerful evidence for evolution that some latter-day creationists have been driven to call it a hoax!

Evidence for the dinosaur-to-bird hypothesis was given a major boost by important fossil discoveries in the 1960s by Yale University's John H. Ostrom. In the 1960s, Ostrom discovered and described a remarkable dinosaur, *Deinonychus antirrhopus*. *Deinonychus* was as far removed as possible from the sluggish and slow dinosaur stereotype. Here was a creature built for speed and equipped with formidable weapons for killing and butchering. The name "*Deinonychus*" means "terrible claw," and the second toe of each foot was modified into a huge slashing claw. Otherwise, the skeleton of *Deinonychus* was virtually indistinguishable from that of *Archaeopteryx* (Benton 2019: 110). It had hollow bones, a fused clavicle (a "wishbone"), and a semilunate carpal in the wrist, as do birds (Benton 2019: 114). We now know that, in addition to its close skeletal similarities, *Deinonychus* was feathered (Benton 2019: 114). A very bird-like dinosaur!

As important as the fossil evidence linking birds to dinosaurs has been the cladistic analyses. Cladistics was developed by German biologist Willi Hennig in the middle of the twentieth century as an effort to make biological systematics more objective. Systematics is the arrangement of groups of organisms into an orderly scheme of classification. Anyone who has taken a high school biology course is familiar with the system of "binomial nomenclature" whereby organisms are given scientific names that specify their genus and species. *Tyrannosaurus* is the genus and *rex* is the species. Genus and species are examples of what biologists call "taxa ('taxon,' singular)," that is, systematic groupings of organisms. Cladistics is also known as "phylogenetic systematics" because it classifies organisms solely in terms of phylogeny, the evolutionary history of species or other taxa, showing their descent from and relations to other groups of organisms.

Cladistic analyses can get rather complex and technical, but the basic idea is simple. Consider any three species, say humans (*Homo sapiens*), cats (*Felis catus*), and boa constrictors (*Boa constrictor*). The three species have many anatomical features in common, such as lungs, hearts, and spinal cords. Yet humans and cats have traits, such as fur (or hair) and mammary glands that boa constrictors do not have. With respect to these three species, we may therefore distinguish between "primitive" traits, like lungs, that all three species share, from "derived" traits, like fur, that humans and cats share, but not boa constrictors. Unlike lungs, then, fur must be an evolutionary novelty that developed after the line leading to humans and cats had split from the line leading to boa constrictors.

Since cats and humans share evolutionary novelties that boa constrictors do not have, then cats and humans must have had a more recent common ancestor—one that possessed those novel traits—than either had with boa constrictors. By considering many different derived traits among many different taxa we may construct phylogenies, representations of the evolutionary relationships between those taxa. The aim of such analyses is to identify "clades," which are monophyletic groups, that is, an ancestral group and *all* and *only* those groups that have the ancestral group as their most recent common ancestor.

According to such analyses, birds are a monophyletic group. All of the greatly diverse orders, families, genera, and species of the class *Aves*, from eagles to ducks to hummingbirds, share a most recent common ancestor that they do not share with any other group of creatures. Extensive and painstaking cladistic analyses have reconstructed avian phylogeny in detail, and indicate that birds evolved from a group of theropods called "coelurosaurs" at some time in the Jurassic (Padian and Chiappe 1997: 75). In fact, just as humans are primates, so birds are dinosaurs:

> Penguins, along with all other birds, are dinosaurs. They are not closely related to dinosaurs, they *are* dinosaurs. Being a dinosaur is a binary condition; there are no degrees of being a dinosaur. Either you are a dinosaur or you're not. Avian dinosaurs, as we call birds, either possess or have ancestors that possessed, all the defining characteristics of dinosaurs. Birds are dinosaurs in the same way that a *Tyrannosaurus*, a *Stegosaurus*, or a *Dreadnoughtus* are.
>
> (Lacovara 2017: 20–21)

Cladistic analysis has been sharply criticized. Various things can confute a cladistic analysis (Mayr 1982: 228–229). For instance, convergent evolution can cause two distantly related taxa to share a common feature as if it were derived from a recent common ancestor. Deciding which traits are phylogenetically informative is not an easy task, and it is addressed by repeated testing with many different characteristics (Benton 2019: 55).

Again, though, the methodological debates may be set aside and left to the experts. The important point here is that a major revision in the way paleontologists think about the relation between dinosaurs and birds seems to have been due to remarkable discoveries and the painstaking application of rigorous methods. To all appearances, the process of convincing (most) paleontologists of the dinosaur/bird connection seems to have been incremental and cumulative. Such an explanation of major conceptual changes certainly lacks the excitement and drama of the Kuhnian or social constructivist accounts. It has only the slight advantage of being true.

A REAL REVOLUTION AND A NON-CONVERSION

The scenario is now familiar: The dinosaurs, having been the lords of creation for over 150 million years, were enjoying one last perfect day, the last day of the Cretaceous. Suddenly, the sky is split by a plummeting object far brighter than the sun. It is an asteroid the size of Manhattan that impacts the earth's crust with a kinetic energy of over ten billion megatons (Benton 2019: 254). The strike occurs in the Caribbean, just off the Yucatan. Massive tsunamis carry devastation far inland on the surrounding landmasses. A vast amount of superheated rock is blasted into ballistic trajectory and rains down all over the earth, making the atmosphere radiate heat like the broiler of an oven, and starting massive fires. After the fires, comes the dark and the cold as the dust of vaporized rock blocks sunlight. Photosynthesis fails, and herbivores starve and then the carnivores that fed on them. The long reign of the dinosaurs abruptly ends, and the survivors, including our mammalian ancestors, find themselves in possession of a world which, without their dinosaurian overlords, invites their proliferation and diversification.

This narrative has become a standard element of scientific accounts and of popular dinosaur lore, featured in PBS programs and even newspaper comics and TV commercials. It is easy to forget that paleontologists still active can remember when such a scenario would have invited ridicule (Benton 2019: 256). Of course, it had been known all along that the dinosaurs became extinct with geological suddenness at the end of the Cretaceous (*geological* suddenness, that is). Popular depictions of dinosaurs as contemporaneous with cave men were, of course, recognized as ludicrously anachronistic. However, from the perspective of human history, geological suddenness can still be a very long time. There were many speculations about the demise of the dinosaurs, but they all assumed that the cause or causes were earth-bound. Most likely, it seemed, changes in climate and other environmental factors made the world less and less viable for dinosaurs—and friendlier to mammalian competitors. In the end, dinosaurs died out in a drier, colder, and more competitive world for which they were not suited.

The acceptance by paleontologists of a sudden—literally sudden—mass extinction that took place at the end of the Cretaceous and the beginning of the Paleogene (called the K/Pg extinctions) therefore once again raises the Kuhnian question: What *really* goes on when scientists change their minds in a fundamental way? Did individual scientists undergo something like a religious conversion, as though, as Kuhn put it, they had been transported to a different planet? Was there a holistic change in their conception of their science? Was there a radical change in standards, values, methods, data, and the meanings of terms? Was there an unavoidable breakdown in communication between proponents of the new extinction theory and defenders of the old-style ones? Did their discourse become incommensurable?

In mid-life Leo Tolstoy famously experienced a profound religious conversion. He reported that after his conversion all the rites and ceremonies of Russian Orthodoxy, which had seemed like meaningless mummery before, now were charged with meaning, while all that had mattered to him in his previous life now seemed unutterably vapid. Does something like this occur when scientists adopt a radical new theory, or is the whole talk about "conversions" overblown and unhelpful in understanding scientific change? Maybe the most fruitful way to approach the question is not from the standpoint of clashing philosophical positions, as was often the case in the "science wars" disputes, but to do Kuhn's job more thoroughly, that is, to look even more closely at apparent instances of theoretical "conversion" in science.

Interestingly, some scientists have self-reported undergoing something like a "conversion" in their views of reigning theories. Perhaps the most prominent was eminent paleontologist David Raup. In his book *The Nemesis Affair* (Raup 1986) he recounts his switch from rejection to acceptance of the "impact" hypothesis, the claim that the K/Pg mass extinctions were due to the cataclysmic impact of a massive extraterrestrial body. In this book Raup recalls his own reactions to this hypothesis, ranging from initial disdainful rejection to acceptance, to, finally, the enthusiastic support of a convert. Do we have here an instance of "conversion" in Kuhn's sense? That is, do we see that Raup not only changed his views about a factual claim—the causes of the K/Pg extinction— but also underwent a fundamental change in basic values, standards, or methodology? Did he come to adopt views that were in any sense "incommensurable" with his previous ones? Raup's case is an excellent case study of alleged "conversions" in science, an opportunity to assess the extent that acceptance of a new paradigm requires a "gestalt switch," in Kuhn's language, a fundamental change of perspective, perhaps even of worldview.

In June 1980, the journal *Science* published the article "Extraterrestrial Cause for the Cretaceous-Tertiary Extinction," authored by Luis Alvarez, Walter Alvarez, Frank Asaro, and Helen Michel (Alvarez, Alvarez, Asaro, Michel 1980). This was the article that precipitated one of the nastiest controversies ever in the earth sciences, one that plumbed depths of acrimony seldom witnessed. Raup's first exposure to the impact hypothesis occurred when *Science* asked him to review the manuscript. Raup reports that his initial reaction was harsh, unduly so he later judged. Yet, according to the timeline in *The Nemesis Affair*, within a very few years Raup, in collaboration with J. John (Jack) Sepkoski, Jr., had begun to publish a particularly radical impact thesis claiming that mass extinctions were periodic and were likely explained by extraterrestrial forces such as comet impacts. The theory of periodic mass extinction soon evolved into the "Nemesis hypothesis," which postulated a dark companion star to the sun, one that had a highly eccentric, 26-million-year orbit that brought it close enough to the solar system that it would gravitationally disrupt the Oort cloud, the solar system's halo of comet nuclei, and send millions of comets cascading into the inner solar system. Some of these, it was thought, would surely hit earth, precipitating mass extinctions.

So, the story that Raup tells makes him look like Paul of Tarsus—one who began as a persecutor and ended up an enthusiastic convert. Do we therefore have here a notable case of a Kuhnian conversion in science? Did Raup suddenly "see the light," abandoning old assumptions wholesale and rushing to a new perspective? Did his view of his discipline or its norms and standards undergo a radical shift? Was the "new" Raup a different kind of scientist from the "old" one, as Augustine and Tolstoy became new beings after their conversions?

The short answer to these questions is: No. Nothing like this happened at all, at least not so far as one can tell from looking at the scientific papers Raup published before, during, and after his supposed conversion. I present a detailed examination of this case in my book *Drawing Out Leviathan* (Parsons 2001) but I can only summarize the results here. True, Raup did in a short period of time come to accept a type of theory that he and most other earth scientists had long dismissed. In 1980 the idea that major changes in the earth's history were catastrophic and extraterrestrial militated against assumptions that had guided geologists for exactly 150 years, since the 1830 publication of the first edition of Volume One of Charles Lyell's *Principles of Geology*. Lyell sought to establish a firm scientific basis for geology by repudiating catastrophism. He argued that geology could not hope to achieve scientific respectability if given license to draw upon *ad hoc* catastrophes to account for any geological puzzle. Only the slow, patient work of showing that earthly effects have earthly causes, causes of the same sorts now observed to operate, could hope to provide hypotheses capable of robust confirmation and rational consensus among earth scientists.

Nineteenth-century philosopher of science William Whewell famously described Lyell's methodological prescriptions "uniformitarianism." Lyell's text is not entirely clear, and there has been debate about what precisely is the sort of "uniformity" Lyell prescribes. However, three theses may be identified as the essence of Lyell's recommendations for geological method:

1. Uniformity of Law: Geological theories must presuppose the same basic laws of nature recognized in other fields of science.
2. Uniformity of Process: Geological theories can postulate only the same kinds of geological processes observed

occurring in the world today, for example, erosion, deposition, uplift, subsidence, or volcanic eruption.
3. Uniformity of Magnitude: Geological processes of unprecedented magnitude must not be postulated. For example, geologists should not invoke floods or volcanic explosions bigger than any previously recorded.

The import of these methodological rules is that geological theories must attempt to explain the features of the earth in terms of processes identical in kind and not exceeding in magnitude those sorts that have been observed. Consequently, geology will generally explain things in terms of very gradual processes working over immense time. Gradualism must be the norm in geology; sudden, massive catastrophes that achieve big results in little time are ruled out. Local catastrophes, such as the earthquake Darwin observed in Chile, are allowable, since they are observed, but disasters of unprecedented magnitude are not to be invoked.

The progress of the earth sciences from 1830 to 1980, which was generally conducted along Lyellian lines (with some exceptions), appeared to justify the gradualist, uniformitarian assumptions. The "impact" hypothesis proposed in the paper by the Alvarezes, Asaro, and Michel, clearly conflicted with the spirit and the letter of the Lyellian strictures. The sudden impact of an extraterrestrial body, causing cataclysmic effects, including mass extinctions accomplished in days or weeks rather than millions of years, clearly offended what Raup calls "the Lyellian paradigm." Worse, as Raup argues in his later book *Extinction: Bad Genes or Bad Luck?* (Raup 1991) theories of sudden mass extinction militate against the Darwinian view that extinction is nothing mysterious or a phenomenon requiring extraordinary causes.

For Darwin, extinction is simply the endpoint of rarity, occurring when there come to be so few members of a species, that a breeding population is no longer maintained. Species which are at a disadvantage vis-à-vis its competitors will become increasingly rare until the extreme of rarity, nonexistence, is reached. Raup calls this the "bad genes" explanation of extinction, i.e., that those organisms that lose in the struggle for existence against better-endowed competitors, will eventually go extinct. This, of course, is the old scenario about the dinosaurs—that they were dimwitted, obsolescent, oversized losers defeated by smarter, faster, fitter mammals. Catastrophic mass extinction theories, on the other hand, propose scenarios which make extinction and survival a matter of luck. When a 10 km diameter asteroid hits your area at a speed of 35,000 miles an hour, unleashing billions of megatons of explosive energy, the fit are obliterated right along with the unfit. Further, the survivors of the worldwide effects of such a catastrophe will be those who just happened to be pre-adapted to the radically changed conditions.

So did Raup and the other practitioners of the earth sciences who came to accept the impact hypothesis undergo a change of paradigms, whereby they cast off their allegiance to the Lyellian and Darwinian orthodoxy, and rushed to embrace a new, neo-catastrophist earth science? No, or rather such a claim would be grossly overstated and simplistic. First, no paradigm, no reigning theory, however entrenched and revered is ever as absolute, monolithic, or despotic as Kuhn thought. Raup mentions several examples of leading and highly respected scientists, who, at the heights of their careers, and long before the *Science* paper of 1980, proposed frankly catastrophic theories, cast in the teeth of the Lyellian proscriptions. For instance: Otto Schindewolf, Professor of Paleontology at the University of Tübingen, proposed in 1962 that the end-Permian mass extinction was caused by the explosion of a nearby supernova. Noted Canadian paleontologist Digby McLaren suggested in 1970 that the mass extinction at the close of the Frasnian stage of the Devonian was due to the impact with an enormous meteoroid. In 1973, Nobel laureate Harold Urey published a paper in *Nature* arguing that several extinction events were caused by impacts with comets (see Raup 1986: 36–42).

These speculations and hypotheses drew little reaction at the time, but the fact that eminent scientists, without endangering their careers, could propose such theories during a time of pervasive anti-catastrophist assumptions, shows that paradigms, and their embedded norms and values, are not omnipotent. Creationists and other scientific outsiders charge that their ideas get no traction in science because reigning paradigms establish inflexible orthodoxies that crush anyone who does not toe the doctrinal line. But such alleged scientific "orthodoxy" is a myth. The reason that creationists get no respect is because their claims are egregious nonsense, not because they have offended an alleged orthodoxy. Within any given science at any given time there exists a considerable theoretical diversity, as well as much disagreement about norms, methods, and values; indeed, this is the standard and usual situation. Science always has its mavericks, dissidents, and eccentrics who reject some or all of received theory, without thereby losing their status as legitimate practitioners. Paradigms, to the extent that they really exist, sit far more lightly on science than Kuhn perceived.

Secondly, and most crucially, when you carefully examine what scientists actually do during a supposed "conversion" episode, you find how inappropriate such language is. Scientists have sometimes reported that they underwent visceral changes in feeling, philosophy, or perspective after accepting a new theory. Yet self-reports of remembered feelings are not terribly reliable. Much more telling is to look at what the scientists *did*, that is, their research and publications before, during, and after, episodes of theory change, to see if we can detect the emergence of fundamental changes in method, technique, values, norms, or practice. Does the "convert" look upon his science or practice it in some profoundly new way or from a radically revised epistemological perspective?

With respect to David Raup, I merely assert here what I argue at length in *Drawing Out Leviathan*: There simply is no evidence that the "pre-conversion" Raup and the "post-conversion" Raup were two different kinds of scientist, or, indeed, that he had in any sense undergone a radical shift in his perceptual, conceptual, or cognitive makeup. With respect to the scientific methods and techniques he employs, the standards he invokes, the values he supports, and, indeed, his *Weltanschauung* in general, the post-impact Raup seems quite compatible with the pre-impact one.

A change of perspective can be psychologically sudden; there are indeed "Aha!" moments. The important question is what preceded the "Aha!" Vague talk about "gestalt switches" and "conversions" tells us nothing. In Raup's case, his motive for changing his mind and accepting a new theory was not that he had any sort of sudden revelation, like St. Paul on the road to Damascus. It seems to have come down to old-fashioned data crunching using established statistical techniques and tests that convinced Sepkoski and Raup of the reality of the 26-million-year periodicity. In other words, to all appearances, Raup seems to have changed his mind because of what the traditional rationalist would recognize as plain old scientific reasons. Radical new theories can be accepted by traditional means.

Clearly, then, Kuhn greatly overstated the extent to which theory-choice criteria are dependent upon paradigms. True, big new theories not only change the way we view the world, but the way we view science. Yet, at any given time, even during "crisis" periods when new paradigms are emerging, there are within scientific communities many *deeply grounded* and *broadly shared* methods, techniques, standards, norms, criteria, and sets of data. If scientists do act irrationally in periods of theory change, if they do stop talking to each other, and if they do resort to *ad hominem* attacks and character assassination—they do not *have* to. The reasons, evidence, and arguments are *there* and are sufficient to make fully rational decisions about theory acceptance or rejection. At no point in the debates over a new theory, however heated they might become, are the participants forced simply to talk past one another or find that their discourse has become incommensurable. As said previously, "Incommensurability," if it is to be at all an interesting notion, must mean more than simply that people can be cantankerous, intransigent, or myopic. Of course they can. To be at all interesting, the charge of incommensurability must mean that, *in principle*, mutually agreeable criteria of theory choice are unavailable, and this circumstance we simply do not find.

Finally, why is it that when I examine what is seemingly a prime instance of "conversion" in science, I find none of the phenomena that, according to Kuhn, always characterize episodes of radical theory change? For instance, I find no evidence that the language of Raup in defending the Nemesis theory was in any sense "incommensurable" with the sort of language the old Raup used. There is simply no reason whatsoever to posit a Tolstoy-like conversion to a whole new way of seeing the world. So, am I wrong or is Kuhn about what happens when scientists change their minds?

Perhaps the differences between us really seem to involve the scale of the changes we are talking about. My study of Raup follows one scientist through a few years of his work. Kuhn, on the other hand, made his reputation as a historian of the Copernican Revolution, and his 1957 book *The Copernican Revolution*, still perhaps the best thing in print on the subject, shows the enormous change in the outlook it engendered. Yet the Copernican Revolution involved many players and took place over a period of approximately 150 years. However, Stephen Toulmin has a wonderful quote in which he talks about how even the largest revolutionary changes in science take place one rational step at a time:

> As his [Kuhn's] historical analysis makes clear, the so-called "Copernican Revolution" took a century and a half to complete, and was argued out every step of the way. The worldview that emerged at the other end of this debate had—it is true—little in common with the earlier pre-Copernican conceptions. Yet, however radical the resulting change in physical and astronomical *ideas and theories*, it was the outcome of a continuing rational discussion and it implied no comparable break in the intellectual *methods* of physics and astronomy. If the men of the Sixteenth and Seventeenth Centuries changed their minds about the structure of the solar system, they were not forced, motivated, or cajoled into doing so; they were given reasons to do so. In a word, they did not have to be converted to Copernican Astronomy; the arguments were there to convince them.
>
> (Toulmin 1972: 105; emphasis in original)

The course of science is often like a thousand-mile journey by foot. We may end up in a place unlike anywhere we have been before, full of marvelous and unexpected sights, yet we get there in the most literally and figuratively pedestrian way—step by step.

How We Know About Big, Complex Things

7

How do we understand big, complex things, like the history of the earth, the history of life, or the earth's atmosphere? At one time, physicists studied simple things with simple apparatus. Galileo discovered the law of falling bodies by rolling balls down an inclined plane. Robert Boyle used a primitive air pump to study the "spring" or compressibility of air. Isaac Newton used a couple of prisms to show that "white" sunlight is a mixture of many colors. Michael Faraday demonstrated the principle of the electric motor using some wire, a battery, a magnet, and a rotating copper rod.

Physicists now do big science with big things—big particle accelerators, big telescopes, and big computers—and now find some of their most interesting and important topics in the sciences of complexity. The physics of climate change is one of those big, important, and complex topics. The confirmation of theories about complex things presents many challenges, yet complex things such as the human mind and body and the environment concern us most immediately and universally. Debates about these issues often become acrimonious when religion, politics, or ideology becomes involved. This chapter will attempt to sort out some of the ways that theories about big, complex things are confirmed.

Previous chapters have considered what we might call "global" challenges to science, that is, fundamental challenges to the

DOI: 10.4324/9781003105817-8

very idea of scientific neutrality and objectivity or the claim that science could give us a true account of natural reality—if there really is such a thing. However, there are also critics who claim to support scientific results, methods, and aims generally, but who regard certain claims normally counted as scientific as overblown, mistaken, or even fraudulent. The researchers who support those claims are likewise derogated as incompetent or worse, and whole fields are dismissed as junk science. Most famously, creationists, both the fundamentalist "young earth" type and their more sophisticated "intelligent design" cousins, have long attacked evolutionary theory and have advocated for the inclusion of their ideas in the public-school curriculum. Occasionally, they would enjoy the support of state legislatures before being shot down by the courts.

Those who claim not to be global science critics, but only skeptics of particular fields or disciplines, might, of course, be hypocrites. You do not really support the aims, methods, and values of science if you abandon them as soon as scientific conclusions go against your creed or threaten your profits. Science does not care about your beliefs or your bottom line. Personal duplicity *per se* is not an issue, but it does become an issue when the disputed science has a significant bearing upon the welfare of others, and when the "skeptics" have enough money and influence to impact public policy.

In fact, hypotheses about complex things can be confirmed so robustly that they can prevail not only over scientific opposition but over the obfuscation and disinformation promulgated by vested interests. Some science really is settled-even science about big, complex things—in the only sense that matters, i.e., that some claims achieve strong and stable consensus among the qualified experts. This chapter will consider climate science and the confirmation of anthropogenic (human-caused) climate change (ACC) as an example of how complex claims are cumulatively confirmed by the intersection of many independent lines of evidence.

The Earth's atmosphere is incredibly complex, and its study must be complex also. Climate is affected by very numerous factors that interact intricately and nonlinearly with multiple feedback loops. The hypotheses of climate science therefore must disentangle and weigh those many complex causal factors, identify confounders, distinguish signal from noise, and avoid that always deadly trap of conflating correlation with cause. Confirming evidence is partial, cumulative, and always accompanied by uncertainties. Yet, over time and with minutely exacting effort, the various uncertainties may be resolved, confounders rooted out and banished, and strong consensus achieved. As with the confirmation of evolutionary theory and plate tectonics, the evidence for ACC has accumulated, conjoined, and converged.

As this book was being written, the United Nations' Intergovernmental Panel on Climate Change released its Sixth Assessment Report, "AR6 Climate Change 2021: The Physical Science Basis." This report, authored by hundreds of scientists in consultation with hundreds more, is alarming to say the least. It reports that a significant degree of global warming will inevitably occur, even if stringent steps are now taken to control it. The report claims that the effects of human-caused climate change are already being felt in blistering heat waves and devastating floods, and that much worse is to come. Life-threatening heat waves and droughts will affect billions worldwide. Yet, they say that the most catastrophic consequences may still be avoided by strong and concerted action.

Despite the appearance of a broad and strong consensus among qualified experts worldwide, disbelievers persist, and are emphatic in disputing the data, conclusions, and predictions of the IPCC scientists. Some of these "skeptics" are more accurately described as dogmatic deniers, being clearly motivated by ideology or vested interest. These may simply be dismissed. The issue has been thoroughly politicized so that your stand on climate change now serves as a litmus test for political loyalties.

In their impeccably researched Merchants of Doubt, Naomi Oreskes and Erik M. Conway have shown just how politics and big money interests have mounted a campaign against climate science (Oreskes and Conway 2011). Billionaires and large corporations, fearing that the warnings of climate science will cut into their profits, have poured vast quantities of cash into the fight to discredit ACC and climate scientists. They have lavishly funded "research" institutes and "think tanks" to spread misinformation and disinformation. Their political allies have harassed and attempted to discredit climate scientists, sometimes making themselves look ridiculous in the process (see Mann and Toles 2016).

However, not all critics are obvious cranks or shills. A critique that appeared as this book was being written was Steven E. Koonin's Unsettled: What Climate Science Tells Us, What It Doesn't, and Why It Matters (Koonin 2021). Koonin is a very distinguished theoretical physicist who has held a number of prestigious academic and governmental positions (though he is not a specialist in climate science). His book has been the center of much lively controversy and has already drawn a number of negative reviews (e.g., Yohe 2021, Boslough 2021). I do not have the space—or the expertise—to hash out the back-and-forth between Koonin and his many critics. However, I will conclude the chapter with some reflections on how to distinguish the genuine skepticism that is a necessary component of science from the kind of ersatz skepticism of those who have an ax to grind and will not be persuaded by any evidence.

To provide context for this discussion of climate science, I will first consider how, in general, we know about big, complex things. I will therefore look at the best-known and clearest example of such a big claim and how it has been abundantly confirmed: Darwinian evolution. Nothing new will be cited here; this history is well-known. I am merely offering the example of evolution to remind us about how we know about big, complex things.

EVOLUTION AND THE HISTORY OF LIFE

How do we account for the diversity of life on this planet? How, for instance, do we account for the existence of 350,000 species of beetles? Do we say, as biologist J.B.S. Haldane mischievously suggested, that God has an inordinate fondness for beetles? How do we explain dinosaurs as long as two city buses and weighing as much as ten elephants? How do we explain a fossil record with successive faunas and floras that over geological time come to look more and more like present ones—a fact staunch opponents of evolutionary hypotheses had recognized before Darwin (Larson 2004: 23)? How do we explain the emergence of a bipedal, big-brained primate that is now, by far, the most numerous large animal on the planet? Since the publication of *On the Origin of Species* in 1859 we have had a theoretical basis for such explanations: evolution by natural selection. How do we know that this explanation is correct?

According to Ernst Mayr (1982: 505–510) Darwin actually offered five independent theories:

1. Evolution as such. The history of life is not static, with organisms always producing "after their kind." New kinds arise by a natural process of descent with modification.

2. Evolution by descent from a common ancestor. All organisms descend from only one original common ancestor, or perhaps a very few original beings. New species arise by a branching process whereby parent species split into descendant species.
3. The gradualness of evolution. The process of evolution is slow, occurring over geological time. There are no sudden "leaps" to new forms.
4. Populational speciation. Speciation occurs by the gradual accumulation of differences between successive populations of organisms. This accumulative process means that descendant species can depart to an indefinite degree from ancestral ones.
5. Natural Selection. The main driver of the process of evolution is natural selection, a process whereby the natural variation of organisms produces some who are "fitter," i.e., more likely to meet the challenges of the ambient environment. These fitter organisms are therefore more likely to survive long enough to pass on their heritable characteristics to offspring.

Here we will consider a very small bit of the evidence for the first, second, and fifth of these claims.

Darwin's mode of argument for each of these theses is chiefly inference to the best explanation (IBE), that is, an appeal to each claim as the most compelling explanation of diverse facts about the natural world. IBE is the mode of reasoning familiar to readers of detective fiction. Sherlock Holmes called his inferences "deductions," but really they were instances of IBE. Typically, Holmes was presented with a mysterious and disturbing incident and a set of puzzling facts, and the solution to the mystery was the best explanation of those facts. It was these seemingly inexplicable facts—and the odder the better—that became the crucial clues when they were recognized not as odd and isolated, but as integral aspects of a criminal scheme.[1]

Charles Darwin, biology's Sherlock Holmes, confronted a much greater mystery than any fictional detective, the "mystery of mysteries," the appearance of new beings on the earth, and his reasoning was like Holmes'. Indeed, Mayr concludes that Darwin's theories did not chiefly rest on new discoveries but were confirmed mostly by " . . . a novel integration of previously known facts" (Mayr 1982: 856).

Consider the fossil record. As mentioned, long before the publication of *Origin of Species*, it had been recognized, even by opponents of evolutionary theories, that there had occurred massive turnovers in organic life over geological time. Trilobites, for instance, flourish in ancient strata and then disappear, never to be found again. Different stages of earth's history were seen to contain their own floras and faunas. These were so distinctive that William Smith, whose work in constructing the canals required him to dig into rock strata across Britain, learned to identify the layers of strata by their characteristic fossils. Confronted by these facts, creationists postulated multiple creations at multiple locales. Darwin postulated a simpler and less *ad hoc* explanation by proposing that as some types of organisms became extinct, new ones developed from ancestral types by a process of descent with modification.

Darwin went out on a limb by appealing to the fossil record. As he realized, that record is very incomplete with many extensive gaps. Fossilization is a very chancy thing, as is the deposition of the sedimentary rocks in which fossils are found. Further, the seemingly sudden appearance of new forms in the rocks appeared to tell against Darwin's claim of gradual evolution and to favor theories of the sudden generation of new forms. Critics noted that the "missing links" between major forms of organisms were indeed missing. Then, in 1861, in the Solnhofen limestones of Bavaria, a remarkable fossil was found. *Archaeopteryx lithographica*, dating from the late Jurassic, is a mosaic of avian features (feathers and a fused clavicle—a wishbone) with many other skeletal features very similar to those of small dinosaurs. In the years since, paleontologists have found many more examples of transitional fossils. The story of the discovery of one of the most remarkable ones is told by Neil Shubin in his book *Your Inner Fish* (Shubin 2008).

One of life's major transitions was the emergence of land animals such as amphibians. Paleontologists have long held that amphibians evolved from fish between 365 and 385 million years ago during the Devonian Period. Yet there were no fossils of transitional creatures. Fish have conical heads with eyes on the side and no neck joints. Amphibians have flat heads with eyes on top and neck joints, permitting them to move their heads without moving their whole bodies. In 2004, Shubin and his colleagues found a remarkable fossil in the Devonian rocks of the Canadian arctic. The creature, given the Inuit name *Tiktaalik*, was a perfect intermediate:

> Fish have scales all over their bodies; land-living animals do not. Also, importantly, fish have fins whereas land-living have limbs with fingers, toes, wrists, and ankles . . . But our new creature broke down the distinction between these two types of animal. Like a fish, it has scales on its back and fins with fin webbing. But, like early land-living

animals, it has a flat head and a neck. And, when we look inside the fin, we see bones that correspond to the upper arm, the forearm, even parts of the wrist. The joints are there too: this is a fish with shoulder, elbow, and wrist joints. All inside a fin with webbing. Virtually all of the features that this creature shares with land-living creatures look very primitive. For example, the shape and various ridges on the fish's upper "arm" bone, the humerus, look part fish and part amphibian. The same is true of the shape of the skull and shoulder.

(Shubin 2008: 22–24)

An equally remarkable set of intermediates documents whale evolution during the Eocene (Thewissen 2014).

Transitional fossils are particularly striking, but the fossil record provided only a small portion of the evidence adduced for evolution by natural selection. Darwin also cited geographical distribution, embryology, and taxonomy. For instance, he noted that organisms on islands closely resembled those on the nearest large land mass but were altered for the conditions of island life, as if by an evolutionary process of adaptation. Taxonomists had long grouped organisms according to morphological similarity, with the most similar placed in the same species, similar species into genera, genera into families, and so on up to the broadest classifications of phylum and kingdom. Darwin proposed that this nesting of types within types suggests a ramifying process of diversification with members of lower taxa having branched from a common ancestor more recently than members of the higher taxa.

As Shubin details, the development of embryology since the time of Darwin provides some striking evidence for the common ancestry of very different organisms such as humans and sharks. Evolution predicts that if presently very different organisms had a common ancestry, we would expect to find deep commonalities beneath the differences, as we have previously noted with respect to skeletal homologies. Such commonalities are also found at the embryonic level. Human and shark embryos each have four arches in corresponding locations that look like gill slits (Shubin 2008: 90). From the first arch jaws develop in both humans and sharks. The cells in the second arch form a bar of cartilage and muscle. In humans this bar becomes bones of the middle ear; in sharks the bar breaks up to form two bones that support the jaws (Shubin 2008: 91). The third and fourth arches supply us with structures for talking and swallowing; in sharks they move the gills (Shubin 2008: 92). These structures are served by four cranial nerves, the trigeminal, the facial, the glossopharyngeal, and the vagus that are present in humans and sharks, exit the brain in the same order, and serve corresponding structures (Shubin 2008: 92). So, as the theory of common descent expects, anatomical differences arise by differential developments from shared embryonic features.

With respect to natural selection, Darwin can argue more than just that this hypothesis is consistent with known facts, but that those facts imply the occurrence of a selection process (Mayr 1982: 479–480). It was known that organisms possess the fertility to reproduce at an exponential rate, but that their populations tend to remain at a fairly constant level, held in check by limitations of the available resources. Darwin noted that without such checks, even the slowest reproducers, such as elephants, would soon fill the world with their numbers. Therefore, there must be severe competition—a struggle for existence—between organisms to determine which will survive to reproduce. Also, there is great variability in populations of organisms so that individuals bear unique sets of traits. Further, many of those traits are inheritable. If then, there is a struggle for existence, and if certain traits make an organism fitter, i.e., better able to compete in that struggle, it follows that survival will not be random, but that probability will favor those with the more adaptive traits. If, for instance, bigger beaks can crack tougher seeds giving a survival edge, even a slight one, succeeding populations will likely have bigger beaks (see Weiner 1995). This unequal and nonrandom survival of differently endowed individuals is the process of natural selection.

When fitter individuals survive to pass on their traits to their offspring, the progeny will inherit those advantages, and so themselves will be more likely to pass these on to their offspring. The consequence is that over successive generations and given the relative stability of the ambient environmental conditions, those advantageous traits will tend to accumulate in successive populations while disadvantageous ones will dwindle. This implies a process of accumulating differences between descendant and ancestral populations, a process that can continue indefinitely. When sufficient differences have accumulated in a descendant population compared to an ancestral one, speciation has occurred.

Of course, I can here only hint at the mass of evidence that, for over 150 years, has been adduced in favor of Darwin's theories (for fuller documentation, see Shubin 2008, Shubin 2020). Enough has been said, though, to indicate some of the features of the evidence that confirmed Darwin's theories:

1. The evidence is *cumulative*. No single observation or bit of evidence provides definitive confirmation. Rather, it is the convergence of many different observations, inferences,

and lines of evidence that coalesce to strongly support Darwin's theories.

2. These theories are highly *consilient*, that is, they provide a common conceptual framework uniting many seemingly unconnected facts and phenomena under a single explanatory scheme. Put simply, many different things can be given the same sort of explanation.
3. The theories generate rigorously testable models. Population genetics uses mathematical models to study the change in gene frequencies in successive populations.
4. The theories have had a very good *track record*. Over time they have had great success in overcoming objections and in accommodating seemingly contrary evidence. Indeed, whole new fields of inquiry, such as molecular genetics, which had the potential to falsify Darwinism, in fact have added evidence supporting it.
5. The causal mechanisms postulated by the theory are consistent with, or even entailed by known facts.

I turn now to the consideration of the development of ACC and the evidence supporting it. Like evolutionary science, ACC is supported by a massive accumulation of evidence. Facts indicative of climate warming and its human causes have piled up in great quantity. New indicators are discovered very frequently. Further, very diverse phenomena are connected to climate change as an underlying cause. When species are found far north of what had always been their range, when glaciers recede at unprecedented rates, and when the spring thaw now occurs three weeks earlier than the old normal, a warming climate is implicated. As noted, ACC has many enemies who have been relentless in their efforts to discredit the science behind it. However, these objections have been addressed in detail. In fact, where the predictions of climate scientists have been most notably wrong is in *underestimating* the speed and extent of climate change. Finally, and perhaps most basically, the mechanisms of climate change are driven by very well-understood and universally accepted physical principles—basic physics and chemistry.

THE RISING TIDE OF CLIMATE CHANGE

Paleoclimatology, the science of ancient climates, tells us that the earth's climate has changed many times over the earth's history. At one time the earth froze over to resemble one of the ice planets of science fiction. At other times the climate was so warm that there was no permanent ice at the poles, and life flourished at extreme latitudes. Climate changes have affected human history. The "Little Ice Age," often dated from the fourteenth into the nineteenth centuries, was a period of regional cooling particularly affecting the North Atlantic region and producing significant and sometimes dire effects for the inhabitants of northern Europe. Natural changes in the earth's orbit and orientation towards the sun, such as the Milankovitch Cycles, have long-term impacts on the earth's climate.

So, the issue is not whether earth's climate has changed and is changing. It is always changing and has been very different at different times in the past. Yet these changes due to natural causes were generally slow, occurring over hundreds or thousands of years. The question now is whether human causes, particularly, the burning of fossil fuels, are causing a rapid change in the earth's climate, one occurring over decades rather than centuries or millennia.

How could this happen? It is basic physics. Solar radiation of all wavelengths strikes the Earth constantly. The Earth warms and radiates the heat in the form of thermal radiation, i.e., infrared radiation of a range of wavelengths. However, not all the radiated thermal energy escapes into space. Rather, certain gases in the atmosphere—"greenhouse gases"—trap that radiation which then heats the atmosphere. Though it is not the most potent greenhouse gas, the most notorious villain here is carbon dioxide, CO_2, which is emitted in stupendous quantities by the burning of fossil fuels. As the concentration of CO_2 and other greenhouse gasses increases in the atmosphere, more heat is retained, and the Earth gets hotter and hotter. Throw in some positive feedback loops, and the effect can be seen to unfold in real time. Some people alive today, like the present author, have seen it in their lifetimes. Those who are in their twenties now may well see dire effects by the time they are in their fifties.

But wait a second. Doomsday scenarios have been graphically portrayed many times before and the world did not end. In 1968 Stanford University biologist Paul R. Ehrlich and his wife Anne Ehrlich authored *The Population Bomb*, which predicted massive famines in the immediate future. The predicted mass starvation did not occur, and the book's critics made it exhibit A for the perils of alarmist prognostication. How do we know that today's climate pessimists are not equally overwrought?

Massive amounts of evidence now support ACC, more than could be summarized in this whole book. Hundreds of different researchers, often unaware of each other, pursued independent lines of evidence that came together to support ACC. Glaciologists noted retreat of glaciers; oceanographers documented the rise in sea levels and temperatures; biologists noted the changes in animal behavior and habitat; climatologists noted climatic warming; physicists and chemists charted the monotonic rise in atmospheric CO_2 over decades—and so on. Taken individually, such facts can often be given various interpretations. However, when confronted with a plethora of facts that conjoin and mutually reinforce to point in a particular direction, and with few or none pointing the other way, it is hard not to get the impression that *something* significant is afoot. However, no set of facts, however impressive, constitutes a science. As Mayr noted, many of the facts Darwin adduced were already known, and evolution by natural selection provided a "novel integration" of these facts. ACC is the theory that provides the novel integration of climate facts.

Note that, as with my sketch of evolution, what follows is not intended to be an original historical exposition or analysis. Rather, drawing on textbook and secondary sources, my aim is merely to indicate how climate science established basic facts and successively met various challenges. The historical account is offered to support the claim made by Gary Yohe in his review of Steven Koonin's *Unsettled*, published in *Scientific American*:

> The science is stronger than ever around findings that speak to the likelihood and consequences of climate impacts, and has been growing stronger for decades. In the early days of research, the uncertainty was wide; but with each subsequent step that uncertainty has narrowed or become better understood. This is how science works, and in the case of climate, the early indications detected and attributed in the 1980s and 1990s, have come true, over and over again and sooner than anticipated.
>
> (Yohe 2021)

From hesitant beginnings, the study of human effects on climate has gained in precision and depth, incorporating a vast range of observations and experimental results. Concomitantly, the projected effects of climate change have become clearer over time as the nature and extent of ACC has been confirmed.

As with evolutionary science, the rancor of opponents of ACC has not stymied progress, but only provided more opportunities to articulate the theory's strength (for a book-length account of the history, see Weart 2008).

Swedish physicist Svante Arrhenius was the first to make a detailed study of the potentially climate-changing effects of increased concentrations of atmospheric CO_2. Arrhenius was not exactly the Darwin of ACC. He was concerned with natural sources of CO_2, not anthropogenic ones. Further, unlike Darwin, whose theories were at least partially accepted by his contemporaries, Arrhenius's hypothesis was quickly subjected to seemingly devastating criticisms. Only after many years was it realized that the critiques were in error and that Arrhenius had been essentially correct. Many hypotheses are ahead of their time, containing deep insights that cannot be generally accepted because the weight of evidence is against them at the time. Then, often many years after one who had proposed the hypothesis is deceased, the hypothesis will be resurrected, and its proposer vindicated. That ultimately triumphant ideas sometimes undergo a long eclipse is not a flaw of science, but an unavoidable consequence of the fact that scientists (or anyone) can do no better than to judge on the basis of the best arguments and evidence available to them.

Arrhenius, a physical chemist, won a Nobel Prize for his demonstration of how salts dissolved in water separate into ions. A man of unusually broad interests, in 1896 he turned his attention to the question that had long stumped geologists: What caused the ice ages? That ice ages had occurred, burying what is now open land under mile-thick sheets of ice, had been an astonishing discovery made earlier in that century. What could cause such dramatic shifts of climate between the frigid ice ages and the warmer interglacial periods (one of which we currently inhabit)?

Arrhenius proposed that fluctuations in the amount of atmospheric CO_2 could initiate feedback loops that would warm or cool the earth:

> . . . a spate of volcanic eruptions might spew out vast quantities of the gas. This would raise the temperature a bit, and that small increment would have an important consequence: the warmer air would hold more moisture. Because water vapor is the truly potent greenhouse gas, the additional humidity would greatly enhance the warming. Conversely, if all volcanic emissions happened to shut down, eventually the CO_2 would be absorbed into soil and ocean water. The cooling air would hold less water vapor. Perhaps the process would spiral into an ice age.
>
> (Weart 2008: 5)

Determining such complex conditions was impossible at the time, so Arrhenius devoted months of painstaking calculation of the direct effects of adding CO_2 to the atmosphere and the consequent variations in water vapor with rising or falling temperature. Warmer air holds more water vapor, so, when relative humidity is held constant, the absolute amount of water vapor will be greater in warmer air. His results were published in his paper "On the Influence of Carbonic Acid in the Air upon the Temperature of the Ground," which Lawrence Krauss rightly described as "groundbreaking" (Krauss 2021: 58). Given the limitations of the available data and the simplifying assumptions he had to make, Arrhenius's achievement was impressive:

> The positive feedback effect of water vapor . . . was incorporated iteratively, keeping relative humidity unchanged in the atmosphere as temperature changed in each successive iteration. In addition to water vapor feedback,

he incorporated the feedback effect of snow cover that retreats poleward as temperature increases, thereby enhancing the warming of the Earth's surface. It is quite impressive that Arrhenius identified two of the most important positive feedback effects and incorporated them into his computation.

(Manabe and Broccoli 2020: 18–19)

The result of his laborious calculations was Arrhenius's estimate that a 50% rise in CO_2 concentration in the atmosphere would increase average global surface temperatures by 3° to 3.5° C and that a doubling of the concentration would cause an average increase of 5° to 6° C. Further, the warming would be greater at the poles than at the equator (Krauss 2021: 62). According to Krauss, Arrhenius proposed the general rule that each time the concentration of CO_2 increased by a factor of 3/2, a change in average surface temperature of about 3° C is expected (Krauss 2021: 62–63). Arrhenius's conclusions seem remarkable prescient. His prediction of a 5° to 6° C rise in surface temperature with a doubling of CO_2 concentration is at the upper end of the sensitivity range of current models (Manabe and Broccoli 2020: 19). At the time, however, these conclusions quickly drew seemingly knock-down critiques.

Swedish physicist Knut Ångström raised two objections, one theoretical and the other experimental. Water vapor is also a potent greenhouse gas, that is, like CO_2 it absorbs thermal radiation from the earth's surface, thereby trapping heat in the atmosphere. According to Ångström's measurement, the frequencies of the radiation trapped by water vapor overlapped the frequencies of the radiation trapped by CO_2. In other words, water vapor traps the same thermal radiation at all the same wavelengths as CO_2, leaving nothing for CO_2 to block. Since water vapor is much more abundant in the atmosphere and a much more powerful greenhouse gas, Ångström concluded that the atmosphere was already effectively opaque to infrared radiation and that adding additional CO_2 to the atmosphere would have a negligible effect (Krauss 2021: 68–69). Further, experimental measurements were made on a 30 cm tube of pure CO_2 that was claimed to have the same amount of CO_2 as a column of air stretching from the ground to the top of the atmosphere. (Only a small percentage of the air is CO_2.) Reducing the amount of CO_2 in the tube seemed to have no appreciable effect on its infrared absorptivity (Krauss 2021: 69).

Arrhenius did overestimate the absorptivity of CO_2 by a factor of approximately 2.5 (Manabe and Broccoli 2020: 19). However, Krauss notes that Ångström's criticisms were also erroneous. More accurate measurements of the spectrum of infrared absorption, not possible in Ångström's day, show that the absorption spectrum of water vapor is highly spiky, with pronounced peaks and valleys, meaning that some infrared wavelengths are strongly absorbed and others only slightly (Krauss 2021: 69). The peaks are narrower and the valleys deeper at high altitudes (Krauss 2021: 70). CO_2, which is well mixed at all levels of the atmosphere, then functions to fill the gaps between the absorption peaks of H_2O vapor, trapping the infrared radiation wavelengths that the water vapor does not and which would otherwise escape into space.

As for the experiment with the CO_2 tube, Krauss says that Angström overlooked an important point:

> What increasing the CO_2 concentration does . . . is increase the *range* of frequencies that can have saturated absorption, and this increases the total rate of IR [infrared] absorption by CO_2 in the atmosphere. So, Angstrom's concern about

saturation was a red herring. Absorption can increase with increasing CO_2 concentration *even if absorption at some frequencies is already saturated*.

(Krauss 2021: 73; emphasis in original)

As more CO_2 is added to the atmosphere, outgoing thermal radiation will contact more CO_2 molecules and more will be blocked across broad wavelengths, thus trapping more heat in the atmosphere.

The upshot is that Arrhenius was indeed correct that increasing the CO_2 concentration in the atmosphere enhances *radiative forcing*, which is the crucial factor in global warming. If the energy coming in from solar radiation equals the energy going out into space from the Earth, then the Earth is in energy balance. However, anything that alters the amount of incoming or outgoing energy disrupts this equilibrium. Eventually, the balance is restored when the Earth heats or cools by the amount necessary to equalize the incoming and outgoing energy. If incoming radiation increases or outgoing radiation decreases, then the Earth will heat until equilibrium is restored. Radiative forcing is the change in energy in minus the change in energy out *before* the earth's temperature has adjusted to restore the equilibrium. Symbolically,

$RF = \Delta(E_{in} - E_{out}) = \Delta E_{in} - \Delta E_{out}$ ("Δ" stands for "change in;" Dessler 2022: 94).

This formula is applicable no matter what the nature of the perturbing factor:

> Radiative forcing is a quantitative measure of how much some perturbation (e.g., an increase in solar constant, increase in in greenhouse gasses) will change the climate.

The advantage of using radiative forcing is that it allows us to express diverse changes to the climate system using a common metric.

(Dessler 2022: 94)

Radiative forcing can be either positive or negative. If positive, it tends to heat the earth; if negative, it tends to cool it.

The amount of radiative forcing due to greenhouse gasses added to the atmosphere since preindustrial times (prior to 1750) can be calculated:

The atmospheric abundance of carbon dioxide increased from 280 to 407 ppm [parts per million] between 1750 and 2018; this corresponds to a radiative forcing of +2.2 W/m^2 [watts of energy per square meter of the earth's surface]. Because of the long lifetime of carbon dioxide in our atmosphere (many centuries), a fraction of the radiative forcing from carbon dioxide the Earth is experiencing right now is due to carbon dioxide emitted in the early 1800s. Increases in methane, nitrous oxide, and halocarbons [also greenhouse gasses] between 1750 and 2018 produced radiative forcings of +0.54, +0.19, and 0.38 W/m^2, respectively . . . Virtually all of these changes in greenhouse gasses were due to human activities. Together, they imposed a radiative forcing of +3.5 W/m^2 between 1750 and 2018. Carbon dioxide contributed about 60% of this, so it was the single most important greenhouse gas emitted by human activities.

(Dessler 2022: 96)

So, basic physics accounts for radiative forcing. However, the amount of global warming cannot simply be read off from radiative forcing due to greenhouse gasses. With something as complex as the earth's atmosphere, things cannot be that simple. Two complications must be considered: aerosols and climate sensitivity.

Aerosols are particles so small and light that they can be suspended in the lower atmosphere for days or weeks until they settle out or are washed out by rainfall. If injected into the stratosphere, for instance by explosive volcanism, they can remain in the atmosphere for several years. Sulfur is emitted from both natural sources and the burning of fossil fuels. In the atmosphere sulfur reacts with other substances to form sulfate aerosols. Sulfate aerosols are highly reflective of sunlight and therefore can lower temperatures by negative radiative forcing. Perhaps the most famous example occurred after the massive eruption of the Indonesian volcano Tambora in 1815, the greatest eruption in recorded history. The vast ejection of volcanic ash added huge amounts of sulfur dioxide to the atmosphere, causing such severe negative radiative forcing that 1816 was known as the "year without a summer." Snow fell in New England and upstate New York in June, and late frosts caused widespread crop failure both in Europe and North America. Food shortages, disease, economic depression, and civil unrest followed in many places.

Dessler summarizes the net effects of all positive and negative radiative forcings:

> Summing all the radiative forcings . . . we get a net radiative forcing of $+2.5\,W/m^2$ between 1750 and 2018. This total comes from positive forcings, primarily the greenhouse gases, partially cancelled by negative radiative forcing, primarily aerosol forcing. Virtually all these radiative forcings are tied to human activities . . . As the planet warms, E_{out} increases . . . Measurements of the Earth's energy balance

show that, for the present day Earth, E_{in} exceeds E_{out} by 0.8 W\m². This means that the planet has warmed up enough in the past 250 years to erase +1.7 W\m² of the radiative forcing. The planet needs to continue warming to erase the remaining radiative forcing—this is warming that we are already committed to and can do little to stop.
(Dessler 2022: 101)

If positive radiative forcing could stop, as by ceasing to add greenhouse gases to the atmosphere, then the climate will eventually stabilize, albeit at a higher temperature than before the industrial era. However, if we continue to add greenhouse gases and increase positive radiative forcing, then the temperature will continue to rise.

"Climate sensitivity" has to do with the Earth's response to radiative forcing, that is, how the Earth warms to restore the energy balance (Dessler 2022: 101). Of particular importance here are positive and negative feedbacks. Positive feedbacks tend to enhance the effect of positive radiative forcing and negative feedbacks work against that effect. "Fast feedbacks" are those that occur rapidly enough to have a significant effect on climate change over the coming century (Dessler 2022: 104). A "slow feedback" is one that takes many years to respond to increasing warmth. For instance, the Greenland and Antarctic ice sheets are so massive that it will take millennia for them to completely respond to global warming (Dessler 2022: 105–106).

An instance of such a rapid feedback effect is the *ice-albedo feedback*. The Earth's albedo is the proportion of solar radiation that is reflected into space. Because ice is highly reflective, more ice on the earth's surface increases albedo and less decreases it. If the Earth heats due to radiative forcing, ice cover will melt, decreasing albedo and less heat will be reflected and more retained. The increasing heat will then melt more ice further decreasing albedo and further increasing warming—and so on. Ice-albedo feedback is therefore a simple example of a positive feedback process.

The strongest positive feedback is the *water vapor feedback*. Warmer air can hold more water vapor and, since water vapor is a powerful greenhouse gas, it causes more warming (As Arrhenius noted!). The strongest negative feedback is *lapse-rate feedback*, which is because warmer air radiates more energy into space. In the ideal case (what physicists call a "black body" emitter), energy radiates from a body in proportion to the *fourth power* of temperature. If radiative forcing therefore causes the upper atmosphere to warm, more power will be radiated into space, partially offsetting the initial warming due to radiative forcing. More ambiguous is cloud feedback. Clouds are highly reflective, so they can both reflect solar radiation away from the Earth or act as a blanket trapping thermal radiation reflected from the ground. The net effect of *cloud feedback* is a matter of controversy, though the consensus is that the overall effect is positive, i.e., that clouds increase warming (Dessler 2022: 103–104).

Of course, determining the net effect of all these factors is difficult. Adding to the complexity is the fact that some factors can act both as radiative forcings and as feedbacks (Dessler 2022: 107). The precise nature of climate sensitivity remains an uncertainty in the understanding of climate change (Dessler 2022: 107). However, uncertainty about the degree of climate sensitivity is far from sufficient to disconfirm ACC or to blunt its warnings. In fact, the history of ACC is one of many successes in overcoming uncertainties and in answering the objections of critics. Here are some examples:

- Worldwide temperatures declined from the 1940s through the 1960s, casting doubt on the projected effects of greenhouse gases (Weart 2008: 116). However, climate scientists noted that during that period, the effects of greenhouse gases had been masked by natural variations and by industrial pollution. Also, the ocean would absorb excess heat before it began to be measurable in the atmosphere (Weart 2008: 116–117). In the 1970s climate scientists began to predict that the continued buildup of greenhouse gases would force the warming signal to stand out from background noise.[2] This began to happen in the 1980s when it was noted that the three hottest years on record had occurred in that decade (Weart 2008: 118). That trend has continued with each decade breaking the heat records of the previous ones.
- Some critics accepted that warming was occurring but denied that it was the effect of greenhouse gases and instead attributed it to fluctuations of solar activity (Weart 2008: 167). Solar fluctuations can affect climate. The "Little Ice Age" occurred during a period of low sunspot activity, indicating a decrease in solar magnetism (Weart 2008: 120). Since the late 1970s, satellites have made accurate measurements of the solar constant (which is not exactly constant), which shows an 11-year cycle with a fluctuation in the solar constant of about 0.1% (Dessler 2022: 114). However, climate does not respond to such high-frequency fluctuations (Dessler 2022: 114). There is no evidence linking solar activity to the rapid rise in global temperatures over the last several decades. Moreover, the critics' argument can be stood on its head: If global climate is that sensitive to small influences, then it must react to the greenhouse gases massively added to the atmosphere (Weart 2008: 168).

- Many critics targeted the computer modeling and the general circulation models (GCMs) used by climate scientists to represent the worldwide circulation of the atmosphere. One charge was that these models failed to account for copious oceanographic information about the temperatures of tropical seas during the last ice age (Weart 2008: 169). These seas appeared to be much warmer than the GCMs could be adjusted to predict. Apparently, the GCMs could not calculate any climates except the present one. However, it turned out that the oceanographic analysis was wrong, and that tropical seas had been significantly colder during the ice age, approximately in agreement with the GCMs (Weart 2008: 169). Apparently, then, the GCMs were so accurate that they could not be manipulated to calculate incorrect temperatures (Weart 2008: 169).
- Some critics claimed that, while CO_2 is increasing in the atmosphere, this is a good thing. Since plants use CO_2 in photosynthesis, more of it will benefit agriculture and forestry (Weart 2008: 170). However, increased CO_2 could also benefit invasive weeds and pests (Weart 2008: 171). Further, plants will reach a point where increased CO_2 fertilization cannot be of extra benefit, while the harmful effects of CO_2 will build.

ACC appears to have the strengths and theoretical virtues enjoyed by evolutionary theory and other well-confirmed theories about big, complex things. Like Darwinian evolution, ACC is supported by a large diversity of facts for which it offers, in Mayr's words, "a novel integration." ACC provides the theoretical model that unites very diverse sorts of data under a shared explanatory framework. It therefore achieves a high degree of consilience. The theory generates rigorously testable hypothetical models, which increase in scope and accuracy over time (see the detailed history of the development of GCMs in Manabe and Broccoli 2020). Further, the theory has a good track record, having met, and overcome seemingly powerful criticisms. Finally, its most important mechanism, radiative forcing, is implied by basic physics. Like natural selection, given certain facts, it *must* occur.

SCIENCE, SKEPTICISM, AND "SKEPTICISM"

Skepticism is essential to science. New ideas, however beautiful or brilliant, must be subjected to the most rigorous vetting that scientific communities can dish out. Scientists criticize bluntly, not because they are cruel, vindictive, or jealous, but simply because most ideas turn out to be wrong, even the ones that sound most plausible. Nature is under no obligation to respect our proposals, however clever they might be. Lacking omniscience, the only way that we can strip away error in the hope of leaving a core that is close to truth is to rigorously reject those ideas that cannot stand up to our best evidence. Skeptical vetting does not guarantee that we will arrive at truth. Maybe with some questions all of our answers are wrong. But *no* answers are worth believing until they have faced down their skeptics.

On the other hand, not all unbelief is skepticism. Skepticism, by definition, is defeasible. That is, when confronted with sufficient evidence, a skeptic will—provisionally—accept the well-confirmed claim. Mere refusal to accept any evidence is not skepticism. As noted earlier, such a person is a dogmatic denier. The doting mother who will not consider that her angel is—as he is—an incorrigible bully and brat, is not a skeptic. Neither are creationists nor anti-vaccine activists. Such persons have motivations for denial that are too deep to be reached by mere reason. However, dogmatic deniers repudiate that label, often with expressions of highest dudgeon. They portray themselves as virtuous skeptics rather than squalid dogmatists. How, then, do we tell genuine skeptics from "skeptics," or at the expense of introducing a neologism, "pseudo-skeptics?" The way forward is shown by that deep philosophical maven, the junior senator from the State of Texas, Ted Cruz.

In a 2015 interview on NPR, Cruz denied that warming was occurring:

> The scientific evidence does not support global warming. For the last 18 years the satellite data—we have satellites that monitor the atmosphere. The satellites that actually measure the temperature showed no significant warming whatsoever.
>
> (McIntyre 2019: 161)

When a layperson, even a U.S. senator, claims to know an important piece of scientific information that is not acknowledged by the scientific community, this is a red flag. Indeed, as Lee McIntyre points out, Cruz could only have based such a claim on an erroneous 2013 IPCC report that had since been corrected, as noted in a *New York Times* article published six months before Cruz's interview (McIntyre 2019: 162, and note 38, p. 242). Perhaps it is unfair to expect a layperson to keep up with the latest developments, but, in that case, he should also not presume to make authoritative pronouncements about the state of the science.

When the interviewer asked if he also questioned other sciences such as evolution, Cruz did not answer directly, but made a claim about the practice of science:

> Any good scientist questions all science. If you show me a scientist who stops questioning science, I'll show you someone who isn't a scientist.
>
> (McIntyre 2019: 161)

Here Cruz conflates the truth that all scientific results are *in principle* tentative, i.e., capable of being revised or rejected in the face of new evidence, with the falsehood that scientific communities do not *in practice* accept certain results as secure. Science simply could not be done if all results were always suspended in doubt. New science builds upon older science, which, at least for the time being, is not under question. If results could never be accepted, any attempt to do new science would be like attempting to build in bottomless quicksand. Indeed, once a theory is broadly accepted it becomes a tool for further exploration. Astrophysics simply could not be done without taking general relativity as a given.

When asked why global warming is so broadly accepted if its credentials are so flimsy, Cruz invokes a conspiracy:

> ... this is liberal politicians who want government power over the economy, the energy sector and every aspect of our lives ... At the end of the day it's not complicated. This is liberal politicians who want government power.
>
> (McIntyre 2019: 161)

When the interviewer implied that perhaps the shoe was on the other foot, and maybe it was the energy companies that were seeking money and power by sponsoring the denial of climate change, Cruz dismissed the imputation as an *"ad hominem"* (McIntyre 2019: 161).

If you say that you are right and that a whole scientific community is wrong, then the onus is on you to say just how this could happen. Are the researchers incompetent or self-deluded? Are they a cabal, promoting a hoax for nefarious purposes? Are they True Believers who concoct pseudoscientific disguises for their ideological infatuations? Some creationists have claimed that evolutionary theory is literally satanic, having been invented by the Prince of Darkness himself. Others, particularly those of the "intelligent design" stripe, are content with charging that evolutionists are in thrall to "materialist" ideology. The recent backlash against COVID vaccines is driven by a farrago of paranoid suspicions and conspiracy theories that invoke power-hungry governments and profit-mad pharmaceutical companies.

Generalizing from the examples offered by Ted Cruz and others, we may note several features that mark invincible pseudo-skepticism and set it apart from the sort of genuine and necessary skepticism:

1. Pseudo-skeptics hardly ever do research in the fields they criticize. If they do have scientific credentials, then those credentials are seldom in that or a related field. Instead of doing research, pseudo-skeptics scour the professional literature to find disagreements among the qualified practitioners. Disagreements are then played up while broad areas of consensus are ignored.
2. Pseudo-skeptics continue to cite data and research after they are out of date— sometimes long out of date. They are also adept at cherry-picking data, that is, citing only data selected to support a favored conclusion. One particularly effective way of presenting biased data is through misleading charts and graphs, as Darrel Huff pointed out in his

1954 classic *How to Lie with Statistics*. For instance, by judiciously depicting just the right parts of a graph you can create an impression opposite of what the whole graph shows.

3. Pseudo-skeptics mischaracterize the practice of science. Ted Cruz was wrong to say that science must question everything all the time. Creationists were wrong to charge that science has sold out to a materialist metaphysic that dogmatically excludes the supernatural. Sandra Harding was wrong to say that the methods of science are incorrigibly corrupted by politics.

4. Pseudo-skeptics propose explanations and causes that are unconfirmed, disconfirmed, or not confirmable. Modern global warming is not caused by fluctuations in the energy emitted by the sun. Autism is not caused by vaccines. Personality traits are not due to astrological influences. Ancient astronauts did not build the pyramids. The diversity of life is not due to acts of supernatural creation performed over a few days 6000 years ago.

5. When challenged to explain how, as they allege, entire scientific communities can be persistently and intransigently wrong, pseudo-skeptics are often forced to propose conspiracies, plots, or nefarious cabals. Thus, according to Senator Cruz, the whole global warming flap is due to the machinations of power-mad liberal politicians, aided by scientific dupes and fellow travelers, all seeking to extend governmental control over all sectors of the economy.

When confronted with the above list, pseudo-skeptics and their defenders are likely to cry "*tu quoque!*" and say that climate scientists have committed many of the same sins. However, like most *tu quoque* arguments, this one has rhetorical bark but no logical bite. Put bluntly, such a riposte is cheap and shallow.

The effort of pseudo-skeptics to project their own faults onto climate science is a transparent ruse. Climate science, like all science, has had its share of missteps, blind spots, and blind alleys. Honest mistakes, corrected in the light of further research, are the norm for science. As Yohe noted in the earlier quote, climate science has progressed from obscurity and uncertainty to clarity and confidence.

Further, the warnings of climate scientists are dire. Will those warnings be heeded? Will climate scientists go down in history as the Cassandras whose prophecies were ignored until doom befell? Climate change threatens to become all four horsemen of the apocalypse—Death, War, Famine, and Pestilence. If Ted Cruz is wrong, he might not live to see the worst of the consequences of the global warming that he denies. However, his children and grandchildren will. And so will yours.

Conclusion
What Is *Really* Wrong with Science

In this book, I have defended what I call the "rationalist" view of science that affirms the existence of a physical universe with an intrinsic and at least partially knowable nature and the efficacy of scientific practice to acquire such knowledge. In the first two chapters, I considered the critiques of these theses by relativists and social constructivists, and have concluded that these arguments are hollow. In the third chapter, I reviewed the original 13 sections of Thomas Kuhn's *Structure of Scientific Revolutions* and concluded that if these are taken as making a case for conceptual relativism or the irrationality of theory choice, they fail to make that case. In the fourth chapter, I considered the flip side of Kuhn, where he affirms elements of scientific rationality, but I still found some grounds for disagreement. In the fifth chapter, I considered examinations of the relation between science and moral and social values as presented by Sandra Harding and Heather Douglas. I found Harding's treatment to be badly flawed and Douglas's to be far more respectable but still problematic. The sixth chapter considered how the much deeper understanding of dinosaurs, how they lived, evolved, and became extinct show how science accumulates knowledge, even during revolutionary episodes. The seventh and final chapter considered the sciences of complexity, how they are confirmed, and the ways to tell genuine skepticism from the fake skepticism of hard-core deniers.

DOI: 10.4324/9781003105817-9

Yet it would be disingenuous to conclude that all is well in the house of science. Psychologist Stuart Ritchie can only be described as a scientific whistleblower, an insider who has exposed all the dirty little secrets. His book *Science Fictions: How Fraud, Bias, Negligence, and Hype Undermine the Search for Truth* is an exposé far more effective than any attempted by Latour and Shapin or any of the other would-be science debunkers (Ritchie 2020). However, debunking is not Ritchie's aim. His intention is to be brutally honest about how science is all too often betrayed by scientists themselves. Sometimes the betrayal is intentional, but often—and even more insidiously—it is unintentional. Only by facing the alarming facts and engaging in unflinching self-examination can scientists take the steps needed to improve science. Making science better is Ritchie's intention.

Ritchie notes that one of the basic assumptions about science, going back to the beginnings of modern science, is that scientific results are open to confirmation or disconfirmation by any qualified practitioner. If a physicist reports certain observations or results, then any qualified physicist, using adequate equipment, should be able to replicate them. On the other hand, if, for instance, a solitary physicist reports an observation and nobody else can see it, then this is a problem. That was what happened in the case of Blondlot and the N-rays. Nobody else could see them, so other physicists were skeptical.

As Ritchie notes, in recent years scientists have been forced to recognize that very many reported results, published in the best journals by the best scientists, could not be replicated. That is, when other scientists, attempted to find the reported effects using other data, they simply could not find the alleged result. Even attempts to reproduce the alleged findings using the exact same data and methods very often would not generate the claimed effects. The failure of replicability occurs across disciplines but is most alarming when it occurs in medical science. In fact, published studies often omit crucial information that would be necessary even to effectively replicate those studies. Most disturbing is the fact that medical treatments have often been based on research that either could not be replicated, or which was reversed by further trials.

Is Ritchie overstating the problem? Isn't it the normal practice of science to overturn earlier results by more stringent and accurate tests? Isn't this how science progresses? Here is Ritchie's reply:

> But scientists have let doctors and patients down by creating such a constant state of flux in the medical literature, running and publishing poor-quality studies that even students in undergraduate classes on research design would recognize as inadequate. Even at the time those original studies were published, we knew how to do better—and yet we didn't.
>
> (Ritchie 2020: 41)

So, how do so many bad studies wind up in print, despite editors and peer reviews? If a journal publishes only 7% of submissions, why does so much junk science still wind up in print? Ritchie identifies four causes: Fraud, bias, negligence, and hype.

By "bias" Ritchie means primarily not some form of bigotry, but bias in the kind of results that scientists want and that journals want to print. People, including scientists, like definite, clear-cut, unambiguous results. We want to know the truth pure and simple, and we do not like to be told when the truth is neither pure nor simple. No matter how carefully the data are gathered, they will usually mix signal with noise. Disentangling signal from noise is, of course, the job of statistics, and that is why statistical analysis is so important in so many different fields. Unfortunately, if misused, statistics can obscure more than clarify.

One of the main ways that statistics can be misused is by what Ritchie calls "p-hacking." "P" stands for probability, specifically the probability that our results are due to an actual correlation between two variables and that the seeming correlation is not accidental—due to random variation introduced by sampling error, say. Suppose, for instance, that you hypothesize that there is a positive correlation between the amount of TV watching and the amount of snack food consumed. That is, those who watch more TV tend to eat more snack food. You run a test, and, sure enough, your results appear to indicate that those who watch more TV tend to eat more snacks. Do your results reveal an actual correlation, or did you just have the bad luck to have gotten a group of people who watch a lot of TV and also happen to eat an unusually large amount of snack food?

For many years, the "gold standard" has been to designate results as statistically significant that have a p-value of .05 or less, that is, that there is only a 5% or smaller probability that you would get your results if there is *not* a real correlation between more TV watching and more snacking. For instance, if your study indicates that those who watch five or more hours of TV a day eat on average 25% more snack food than those who watch TV three hours or less, are these results significant, or have you just had the bad luck to have selected persons who are atypical in their snacking habits? If your results have a p-value of .05 or less, send off your study for publication, and wait for the accolades to pour in!

And that is the problem. Everybody wants to find something positive—scientists, editors, and peer reviewers. Nobody wants null results. So, what do you do if you keep getting null results? You "file drawer" them, that is, you put them aside and keep looking. However, just as nature is not obligated to respect our moral values, so it is not obligated to respect our career ambitions. We need null results. We might prefer positive results for our hypotheses, but nature might not, and if we want to know what *nature* says about our hypotheses, we cannot ignore null results.

So, what do you do if you keep coming up with null results? Do you abandon your hypothesis? No, you p-hack! According to Ritchie (Ritchie 2020: 99) there are two main ways to p-hack, that is, to massage your data until something statistically significant emerges. One way is to simply keep reanalyzing the data, looking at it in all sorts of different ways, until, by chance, some statistically significant result emerges, and that is the one you send off for publication. The other mode of p-hacking is to take a set of data and just throw at it every statistical test you can think of until some statistically significant result is found, and you then declare that this is what you were looking for all along! Ritchie amusingly calls this the "Texas sharpshooter" trick, whereby you shoot random holes in the wall and then draw bullseyes around the holes to show what a great shot you are. Believe it or not, lots of studies p-hacked in these ways make it into the most reputable journals.

So, how does p-hacking prevent replication? As Ritchie explains (Ritchie 2020: 107), the problem is that p-hacked studies "overfit" the data. If you try hard enough you can often find a statistical analysis that fits a given body of data quite well, maybe very well. The problem is that such an analysis cannot be extrapolated to other sets of data, and so the study cannot be replicated. Science seeks *general* knowledge, that is, knowledge that can usefully be applied in any relevant context. It is not scientifically interesting to know that in July 2016, 20 random adults in Hoboken, NJ who watched a lot of TV also ate a lot of snack food. What we want to know is whether, in general, those who watch more TV tend to eat more snack foods—not just in Hoboken, but everywhere. That is why studies that cannot be replicated are useless.

In addition to bias, Ritchie documents how "negligence, fraud, and hype" have distorted science. In general, why is science susceptible to such distortions? Ritchie's answer is what he calls "perverse incentives." An incentive becomes perverse when it rewards bad behavior and punishes good behavior. Politics, for instance, contains many perverse incentives. Those in elected office are very strongly motivated to keep themselves and members of their party in office. Because this incentive is so strong, both Democrats and Republicans have resorted to gerrymandering whenever they could. To gerrymander is to draw district boundaries in such a way as to keep your voters in the majority. Gerrymandering, in short, is the reverse of democracy. In a democracy, the voters choose their representatives; under gerrymandering, the representatives choose their voters.

So, politics has many perverse incentives that reward bad behavior, but so does science. The high-pressure, ultra-competitive system we now have was supposed to reward good science and weed out the bad, but, perversely, it all too often does the opposite. In my view, the basis for perverse incentives in science comes down to two evils: The tyranny of money and the tyranny of metrics.

In my academic career, I have seen the emergence of the "bottom line" mentality in academe, which is a fixation on generating revenue by increasing enrollment, enlarging class sizes, and replacing full-time faculty with adjuncts. The bottom-line mentality has hit the sciences with a vengeance. Scientific research is expensive, and it is now mostly paid for by grants. To get a grant, the researchers need publications and the more the better because the number of publications is one criterion that grant committees weigh very heavily.

Universities also pressure their researchers to publish copiously. When scientists come up for tenure, a main consideration is the number of publications. Further, universities themselves are ranked both officially by governments and unofficially by many organizations such as the *U.S. News and World Report* rankings. Government rankings of universities can determine the amount of taxpayers' money allotted. The publication rate of the faculty is, of course, one major way that schools are ranked. The consequence is that scientists are under enormous pressure to publish, and it is the numbers that matter, not the more difficult to assess qualities that make good science, like the use of rigorous methods and replicability. "Just crank 'em out" is the message.

This high-pressure, hypercompetitive system therefore incentivizes quantity over quality. Small wonder, then, that scientists will often cut corners, fail to exercise due diligence, submit only positive results for publication, and hype their findings to one and all. Further, since scientists are not dumb, they figure out ingenious ways to game the system. One method is amusingly called "salami slicing." Just as your deli can slice the salami thick or thin, so can papers be sliced up into bits and separately published, thereby inflating the publication count. Thus, one 12-page paper can be diced up into four three-page papers. Your C.V. gets four new lines instead of one! The grant committee or your dean will be four times as impressed!

When the paper itself becomes the product, the search for knowledge must take a back seat. The scientific paper, which was supposed to be the means whereby new knowledge is communicated to the world, becomes the end in itself. Putting science on a business model therefore incentivizes the various behaviors that make for bad science. The lesson, which should have been obvious all along, is that knowledge cannot be manufactured like widgets. Progress in science is sometimes rapid and sometimes agonizingly slow. We cannot dictate the pace of discovery. We are not in control. Nature cannot be put onto a timetable and made to deliver knowledge on demand. Nature does not give a damn whether you get tenure or a grant.

To get good science, you do not have to put scientists into a pressure cooker. Past scientists, including some of the greatest ones, would probably never get tenure or grants today. Charles Darwin got his great idea of natural selection and then waited twenty years to publish it, writing a two-volume work on barnacles in the meantime. Isaac Newton spent far more time practicing alchemy and writing about biblical prophecy than doing physics. Albert Einstein was considered a goof-off and had to work in a patent office until, at age 26, he published three papers, any one of which could have won him the Nobel Prize (and one did). Today's bureaucracy does not have time to wait on a Darwin, Newton, or Einstein to move at his own pace. They need measurable, quantifiable, objective results NOW. To get the numbers they need, they turn to metrics.

A metric is a numerical measure. You take something previously regarded as intangible and you put a number on it. The most famous metric is probably I.Q. Something as variegated, quirky, and idiosyncratic as human intelligence is reduced to a single number. When metrics are adopted, they take the place of judgments based on human intuition and insight, the sort of practical wisdom that can only come through experience. The ostensible advantage of metrics is that they are objective, transparent, and accessible in a way that the murky judgments of human expertise are not. Metrics give you a number, and numbers project an air of authority. It is these seeming advantages of metrics that have made their employment the obsession of all those in managerial positions who crave tools that give clear authority to their assessments, and their consequent apportioning of praise, blame, jobs, and money. Metrics have therefore become the fixation of CEOs, bureaucrats, and university administrators. (For a potent critique of the metrics madness, see Muller 2018.)

Metrics have been adopted to supposedly measure the quality of scientific work. One such metric is to count the number of times a scientific paper is cited by other scientists. At a somewhat more sophisticated level is the h-index. The h-index is determined by the number of papers a scientist has written that have been cited at least that number of times. Thus, if a scientist has written 33 papers that have each been cited at least 33 times, then his or her h-index is 33.

One problem with metrics is that they can be gamed too. If citations are what matters, how do you increase your citations? You cite yourself, of course! This is like increasing the number of "likes" on Facebook by liking yourself a lot. If this seems a bit obvious, then when you peer review another paper, you can recommend that it cite your papers. Another trick is to recycle major parts of your old papers into "new" papers and get them published.

As Ritchie notes (Ritchie 2020: 192), the hard lesson here is the one expressed in "Goodhart's Law": "When a measure becomes the target, it ceases to be a good measure." In other words, if the goodies go to those whose works are heavily cited, then getting citations becomes the main goal. In that case, the number of citations, which is supposed to be a measure of scientific quality, becomes a measure of how avidly people seek citations, and so the whole point is lost.

My own reflection on reading Ritchie is an impression of irony. All sorts of ideological enemies have attacked science from the left and from the right. Big money interests have tried to undermine science when its results threatened their profits. However, as I have attempted to show in this book, the enemies of science can be met with powerful arguments, and the emptiness of their critiques exposed. Ritchie, however, shows that science is its own worst enemy. It is scientific culture that incentivizes quantity over quality, hype over humility, and negligence over diligence.

In his final chapter, Ritchie offers his suggestions about how to make science better. So, science's own worst enemies are all too often scientists themselves. Sometimes, as in cases of fraud, the moral failures of individual scientists are to blame for junk science. As with racism, however, the basic problem is not individual but systemic. The system must change so that perverse incentives will be eliminated and doing the right thing will be incentivized. Ritchie's recommendations can be summarized by three headings: Accountability, transparency, and humility.

These three are interconnected. Greater transparency will lead to greater accountability. In the present system scientists often jealously guard their data and give only sketchy accounts of their methods and procedures. By preregistering trials, other scientists will know ahead of time what the aim of the study is and how it is to be carried out. Preprinting results will allow human and automatic checking of data and statistics as well as methods, analyses, and inferences. More openness in the whole investigative process, from conception to results, will greatly reduce p-hacking, correct for negligence, and make it easier to detect and punish fraud. For instance, the preregistration of trials will not permit researchers to be "Texas sharpshooters," whereby statistical tests are run until something statistically significant falls out, which is then declared the purpose of the study. Preregistration will state ahead of time the effect that is sought and how it is to be tested.

There must be broader cultural changes in how science is pursued and what kinds of behavior is rewarded. Journals must be willing to publish null results and replication studies as well as those reporting new findings. In general, the obsession with findings that are "new," "exciting," and "groundbreaking," must be replaced with an emphasis on science that is solid, if more boring. As Ritchie asks at one point, what is the point of breaking new ground if you never build anything on it? Scientists will be less tempted to hype their work if it is recognized that not every worthwhile result has to be a "Eureka!" moment. As Thomas Kuhn observed, most workaday science is "puzzle solving," that is, not revolutionary breakthroughs, but the slow, steady, cumulative extension of knowledge.

Universities can help by rewarding with hiring and promotion of those who display a commitment to openness and transparency in their research. In general, there must be a shift away from judgments based on quantity and towards those based on quality. This will be difficult to achieve since quality is so much harder to assess than quantity. Anybody can count the number of publications on a C.V. It is much harder to assess whether those publications really contribute to our collective understanding of nature. Metrics that supposedly give more objective measures of quality can all be gamed. "Salami slicing" and self-plagiarism can artificially inflate the number of publications. Self-citation can make a paper appear much more influential than it is. Speaking for myself and not Ritchie, it seems to me that there must be less reliance upon spurious and corruptible metrics, and more reliance on professional judgment.

Finally, Ritchie recommends greater humility on the part of scientists. I think it would be better to say that there needs to be more of a sense of community and less one of competition. It needs to be recognized that science is a genuinely collective, communal pursuit. It takes a village to understand the world. Those who criticize your work should not be seen as enemies, but as cooperators working with you to do better science. Those who save you from error should be thanked, even if they express themselves bluntly. The highly individualistic and egotistical quest for scientific glory needs to be replaced by a shared pride. Look what WE have done, not what I have done. Science is the glory of our species, and we should collectively take pride in its accomplishments.

Notes

1

1. Quine's arguments about underdetermination were developed overtime and expressed differently in different loci. Supporters say his views "evolved," and critics say he "waffled." (For a thorough account of the twists and turns in Quine's views on these topics, see Zammito 2004: 25–33.) What I present here is not intended as an exegesis of Quine, but a synthes is or reconstruction of what Quine and others have frequently said about underdetermination. That is, I am not attempting to state precisely what Quine or any other given thinker said, but to present the underdetermination thesis in what I consider the clearest and most cogent way.
2. Another argument of Quine's often adduced by relativists is the "indeterminacy of translation (IT)." Imagine that Englishman Robinson Crusoe is trying to learn his native friend Friday's language. While strolling through the brush, they are startled as a creature runs suddenly across their path. "Rabbit!" says Crusoe; "Gavagai!" says Friday. So, Crusoe naturally thinks that by "gavagai" Friday means "rabbit," and that is how he translates "gavagai" into English. However, in the beliefs of Friday's tribe, what speakers of English call rabbits are regarded as reincarnated ancestors. Therefore, everytime Crusoe has the sensory stimulation that makes him say "rabbit," Friday, responding to his own sensory stimuli, will say "gavagai," yet Crusoe and Friday mean something entirely different by those terms. They may never realize that though their use of these terms is precisely coordinated in response to shared occasions of sensory stimulation, each means something entirely different.

    The same would apply even to speakers of the same language. How can I know that someone else means just what I mean by "rabbit?" Allegedly, our only evidence for what the other means is that we use the same term on the same occasions. There is no way to get into another's head to grasp his or her rabbit concept. Therefore, relativists could claim, if we cannot even know what others mean by their terms, but only experience a consistent coordination of stimuli and utterance, how can we compare our different concepts to argue that ours are the objectively true ones?

One problem here is that the IT contradicts the UT. The UT assumes that there can be indefinitely many theories of a shared set of facts or data. However, if two speakers, even of the same language, can never know that they are talking about the same thing when they use the same words, which IT implies, then there can be no notion of a set of acknowledged facts shared between them. Each speaker's understanding of "the facts" will be determined by that speaker's own theory, that is, each theory incorporates its own "facts." Theories therefore are not underdetermined. Proponents of different theories simply talk past each other.

2

1. Astronomers use the term "magnitude" instead of "brightness" and employ a logarithmic scale of magnitude such that a star of first magnitude is 2.512 times as bright as one of second magnitude, which is 2.512 times as bright as one of third magnitude, and so on. "Relative magnitude" is how bright the star appears to us. "Absolute magnitude" is how bright the star would appear at a standard distance of ten parsecs (32.6 light years). "Luminosity" is not quite the same thing as magnitude. Luminosity is the total amount of energy emitted by a star, summed up over all wavelengths.

3

1. This is a private communication from my colleague Paul Wagner:

    I had lunch several times with Kuhn the year I was in Cambridge MA. One time I showed a reprint of his black-body radiation book. He wrote a note in it saying this was his best book. It surprised me. He said "Everyone talks about *Structure*. I have been plagued with things I quickly threw together for that book." And then he went on to say, "I am a physicist for gosh sakes. Of course I think predicates of true/false and right/wrong are meaningful!"

And then he went on to say a combination of careless writing and an audience that thought this gave them sophisticated entré to the world of serious science, spun his talk of paradigms and incommensurability into a cause to be championed far more than he ever intended or even thought. All this set him to work on a book in philosophy of language which I do not believe he ever got to finish.

He also said - but he has published this as well - of course Aristotle and Newton could have cloud chamber apparatus explained to them. They were smart. They could eventually understand the reasoning behind naming different tracings as particles just as we can understand much of what they were trying to do as we come to understand more of their theoretical platform. Most of theirs is available for discovery by historians and philosophers. Ours is in the process of being built but that doesn't mean geniuses like Newton and Aristotle couldn't grasp it despite instruction.

The Kuhn quote is from Wagner, P. and Fair, F., *Education for Knowing*. Rowman & Littlefield, NY, 2020, pp. 5–6. Frankly, assuming Professor Wagner's recollections are correct (which I do), I am left nonplussed, or maybe "gobsmacked" would be more accurate. If this expresses Kuhn's real beliefs, then I can only conclude that he would have fully agreed with nearly all of my criticisms of *Structure* presented in this chapter.

2. In a passage from *Origin of Species* quoted by Kuhn (150), Darwin admits that he probably will not be able to convince established naturalists whose minds are stocked with facts long viewed from a viewpoint opposite of his, and he places hopes in the rise of young naturalists who will be capable of greater impartiality. Again, however, intransigence is not incommensurability. Reluctance to consider a new theory, or even deep psychological inhibition, does not imply the sort of in-principle semantic breakdown that incommensurability must be if it is to be of philosophical interest.

3. For information about Barbara McClintock and "jumping genes," see "Barbara McClintock and the discovery of jumping genes," by Sandeep

Ravindran, in PNAS December 11, 2012, 109 (50) 20198–20199. https://www.pnas.org/content/109/50/20198

For a good popular account of the discovery of dark energy, see Tyson, Neil deGrasse, *Astrophysics for People in a Hurry*. New York: W, W. Norton & Company: 94–114).

5

1. From 1991 to 1996, Heather Douglas and I were graduate students in the Department of History and Philosophy of Science at the University of Pittsburgh. I have always deeply respected her as a person and a scholar. Any criticisms I offer should not be taken as in anyway questioning the importance and integrity of her work.
2. Pinker lists 16 such critics, 2002: 342.

7

1. In the story, "The Adventure of the Speckled Band," a distraught young woman, Helen Stoner, comes to Holmes and tells him of the death of her twin sister two years earlier, who died mysteriously just before her planned wedding. As she was dying she cried out the mysterious words "The speckled band!" Now, sleeping in the same bedroom where her sister died, and herself soon to be wed, Ms. Stoner has begun to hear strange noises and has become fearful for her own life. Holmes investigates and uncovers a number of odd facts:

   1. Ms. Stoner's bed has been bolted to the floor and cannot be moved.
   2. Next to the bed is what appears to be a bell cord, but it is not attached to any bell.
   3. A ventilator shaft has been installed connecting Ms. Stoner's bedroom with the bedroom of her stepfather.

   Holmes has also encountered Ms. Stoner's stepfather, Dr. Grimesby Roylott, a man of violent character and evil disposition, who has a financial interest in preventing the marriage of his stepdaughters. Dr. Roylott served in India and keeps dangerous Indian animals as pets.

   From these facts, Holmes infers that the "speckled band" was a venomous Indian snake, sent from Dr. Roylott's room to Ms. Stoner's bedroom through the ventilator shaft and down the fake bell cord to attack

the sleeping woman in the immovable bed. This, of course, turns out to be the solution to the mystery. Holmes' genius was in recognizing that only the hypothesis of a bizarre plot involving a poisonous serpent would explain each of the three odd and seemingly unrelated facts listed above. No alternative hypothesis even begins to match the explanatory power of the serpent hypothesis. Finally, the hypothesis is congruent with everything else we know, particularly the character and motivations of Dr. Roylott.
2. See Walter S. Broecker, https://www.science.org/doi/10.1126/science.189.4201.460.

Bibliography

Alvarez, L.W., Alvarez, W., Asaro, F, and Michel, H.V., "Extraterrestrial Cause For the Cretaceous-Tertiary Mass Extinction," *Science*, 208: 1095–1108.[Q1]

American Institute of Physics, "Data on Women in Physics and Astronomy," https://www.aip.org/statistics/reports/women-physics-and-astronomy-2019

Asimov, I., *The Universe: From Flat Earth to Quasar*. New York: Walker and Company, 1966.

Asimov, I., *Understanding Physics*. New York: Barnes & Noble, 1993.

Benton, M.J., *The Dinosaurs Rediscovered: How a Scientific Revolution Is Rewriting History*. London: Thames & Hudson, 2019.

Bernstein, R.J., *Beyond Objectivism and Relativism: Science Hermeneutics, and Praxis*. Philadelphia, University of Pennsylvania Press, 1983.

Bird, A., *Thomas Kuhn*. Princeton: Princeton University Press, 2000.

Blackburn, S., *The Oxford Dictionary of Philosophy*, 2nd edition. Oxford: Oxford University Press, 2005.

Boghossian, P., "What the Sokal Hoax Ought to Teach Us," in *The Sokal Hoax: The Sham that Shook the Academy*, Franca, L. (eds.), Lincoln, NE: University of Nebraska Press, 2000: 172–182.

Boghossian, P., *Fear of Knowledge: Against Relativism and Constructivism*. Oxford: Oxford University Press, 2006.

Boslough, M., "A Critical Review of Steven Koonin's 'Unsettled,'" *Yale Climate Connections Newsletter*, May 25, 2021: https://yaleclimateconnections.org/2021/05/a-critical-review-of-Steven-Koonins-unsettled/

Brown, M.J., "The Descriptive, the Normative, and the Entanglement of Values in Science," in *The Rightful Place of Science: Science, Values, and Democracy*, Richards, T. (ed.), Tempe, AZ: The Consortium for Science, Policy and Democracy., 2021: 51:65.

Brusatte, S., *The Rise and Fall of the Dinosaurs: A New History of a Lost World*. New York: Harper Collins, 2018.

Carpenter, K., "A Closer Look at the Hypothesis of Scavenging versus Predation by *Tyrannosaurus rex*," in *Tyrannosaurid Paleobiology*, Parrish, J.M., Molnar, R.E., Currie, P.J., and Koppelhus, E.B. (eds.), Bloomington, IN: Indiana University Press, 2013: 265–277.

Chalmers, A.F., *Science and Its Fabrication*. Minneapolis: University of Minnesota Press, 1990.

Chalmers, A.F., *What Is This Thing Called Science?*, 4th edition. Indianapolis: Hackett Publishing Company, 2013.

Clark, D.H. and Clark, M.D., *Measuring the Cosmos: How Scientists Discovered the Dimensions of the Universe*. New Brunswick, NJ: Rutgers University Press, 2004.

Desmond, A.J., *Ancestors and Archetypes: Paleontology in Victorian London, 1850–1875*. Chicago: University of Chicago Press, 1982.

Dessler, A.E., *Introduction to Modern Climate Change*, 3rd edition, Cambridge: Cambridge University Press, 2022.

Donoghue, M.J., "Homology," in *Keywords in Evolutionary Biology*, Keller, E.F. and Lloyd, E.A. (eds.), Cambridge, MA: Harvard University Press, 1992.

Douglas, H., *Science Policy and the Value-Free Ideal*. Pittsburgh: The University of Pittsburgh Press, 2009.

Douglas, H., *The Rightful Place of Science: Science, Values, and Democracy*, Richards, T. (ed.), Tempe, AZ: The Consortium for Science, Policy and Democracy, 2021.

Fastovsky, D.E., "Ideas in Dinosaur Paleontology: Resonating to Social and Political Context," in *The Paleobiological Revolution: Essays on the Growth of Modern Paleontology*, Sepkoski, D. and Ruse, M. (eds.), Chicago: Chicago University Press, 2009: 239–253.

Feynman, R., *The Pleasure of Finding Things Out*, Robbins, J. (ed.), New York: Basic Books, 2005.

Fish, S., "Professor Sokal's Bad Joke," in *The Sokal Hoax: The Sham that Shook the Academy*, Franca, L. (eds.), Lincoln, NE: University of Nebraska Press, 2000: 81–84.

Galileo, G. *Siderius Nuncius or the Sidereal Messenger*, Van Helden, A. (ed.), Chicago: University of Chicago Press, 1989.

Gimbutas, M. *The Civilization of the Goddess: The World of Old Europe*. San Francisco: HarperSanFrancisco, 1991.

Gould, S.J. *The Mismeasure of Man*. New York: W.W. Norton, 1981.

Gribbin, J., *Science: A History 1543–2001*. London: Penguin Group, 2002.

Gross, P.R. and Levitt, N. *Higher Superstition: The Academic Left and Its Quarrels with Science*. Baltimore: The Johns Hopkins University Press, 1994.

Grossman, L., "Pluto's Place," *Science News*, 200, #4: 20–23.[Q2]

Haack, S., *Defending Science—Within Reason: Between Scientism and Cynicism*. Amherst, NY: Prometheus Books, 2003.

Hacking, Ian, "Introductory Essay," in *The Structure of Scientific Revolutions*, Kuhn, T. (ed.), 4th Edition, Chicago: University of Chicago Press, 2012.

Harding, S., *Whose Science? Whose Knowledge? Thinking from Women's Lives*. Ithaca, NY: Cornell University Press, 1991.

Horner, J.R. and Lessem, D., *The Complete T. rex*. Simon & Schuster, 1993.

Hull, D.L., *Darwin and His Critics: The Reception of Darwin's Theory of Evolution by the Scientific Community*. Chicago: The University of Chicago Press, 1973.

Hurlburt, G.R., Ridgely, R.C., and Witmer, L.M., "Relative Brain Size and Cerebrum in Tyrannosaurids: An Analysis Using Brain-Endocast Quantitative Relationships in Extant Alligators," in *Tyrannosaurid Paleobiology*, Parrish, J.M., Molnar, R.E., Currie, P.J., and Koppelhus, E.B. (eds.), Bloomington, IN: Indiana University Press, 2013: 135–154.

Johnson, G., *Miss Leavitt's Stars: The Untold Story of the Woman Who Discovered How to Measure the Universe*. New York: W.W. Norton, 2005.

Johnson, P.E., *Darwin on Trial*. Washington DC: Regnery Gateway, 1991.

Khalidi, M.A., "Incommensurability," in *A Companion to the Philosophy of Science*, Newton-Smith, W.H. (ed.), Oxford: Blackwell Publishers, 2000: 172–180.

Kirk, R., *Relativism and Reality: A Contemporary Introduction*. London: Routledge, 1999.

Kitcher, P., *Abusing Science: The Case against Creationism*. Cambridge, MA: MIT Press, 1982.

Kitcher, P., *Living With Darwin: Evolution, Design, and the Future of Faith*. Oxford: Oxford University Press, 2009.

Klein, E.R., *Feminism under Fire*. Amherst, NY: Prometheus Books, 1996.

Koonin, S., *Unsettled: What Climate Science Tells Us; What It Doesn't; And Why It Matters*. Dallas, TX: BenBella Books, 2021.

Krauss, D.A. and Robinson, J. M., "The Biomechanics of a Plausible Hunting Strategy for *Tyrannosaurus rex*," in *Tyrannosaurid Paleobiology*, Parrish, J.M., Molnar, R.E., Currie, P.J., and Koppelhus, E.B. (eds.), Bloomington, IN: Indiana University Press, 2013: 251–262.

Krauss, L.M., *The Physics of Climate Change*. New York: Post Hill Press, 2021.

Kuhn, T., "Objectivity, Value Judgment, and Theory Chorice," in *The Essential Tension: Selected Study in Scientific Tradition and Change*, Chicago: University of Chicago Press, 1977.

Kuhn, T., "Afterwords," in *World Changes: Thomas Kuhn and the Nature of Science*, Horwich, P. (ed.), Chicago: University of Chicago Press, 1993: 311–341.

Kuhn, T., *The Road Since Structure*, Conant, J. and Haugeland, J. (eds.), Chicago: University of Chicago Press, 2000.

Kuhn, T., *The Structure of Scientific Revolutions*, 4th Edition. Chicago: University of Chicago Press, 2012.

Lacovara, K., *Why Dinosaurs Matter*. New York, Simon & Schuster, 2017.

Lakatos, I. and Musgrave, A., *Criticism and the Growth of Knowledge*. Cambridge: Cambridge University Press, 1970.

Larson, E.J., *Evolution: The Remarkable History of a Scientific Theory*. New York: Random House, 2004.

Latour, B., *Science in Action: How to Follow Scientists and Engineers through Society*. Cambridge, MA: Harvard University Press, 1987.

Latour, B. and Woolgar, S., *Laboratory Life: The Construction of Scientific Facts*. Princeton: Princeton University Press, 1979.

Laudan, L., "A Critique of Underdetermination," in *Scientific Inquiry: Readings in the Philosophy of Science*, Klee, R. (Ed.), Oxford: Oxford University Press, 1999.

Manabe, S. and Broccoli, A.J., *Beyond Global Warming: How Numerical Models Revealed the Secrets of Climate Change*. Princeton: Princeton University Press, 2020.

Mann, M.E. and Toles, T., *The Madhouse Effect: How Climate Change Denial Is Threatening Our Planet, Destroying Our Politics, and Driving Us Crazy*. New York: Columbia University Press, 2016.

Marcum, J.A., *Thomas Kuhn's Revolutions: A Historical and an Evolutionary Philosophy of Science?*, London: Bloomsbury Academic, 2015.

Marsonet, M., *Science, Reality, and Language*. Albany, NY: State University of New York Press, 1995.

Mayr, E., *The Growth of Biological Thought: Diversity, Evolution, and Inheritance*. Cambridge, MA: The Belknap Press of Harvard University Press, 1982.

Mayr, E., *One Long Argument: Charles Darwin and the Genesis of Modern Evolutionary Thought*. Cambridge, MA: Harvard University Press, 1991.

McIntyre, L., *The Scientific Attitude: Defending Science from Denial, Fraud, and Pseudoscience*. Cambridge, MA: The MIT Press, 2019.

Mitchell, W.J.T., *The Last Dinosaur Book: The Life and Times of a Cultural Icon*. Chicago: University of Chicago Press, 1998.

Molnar, R.E., "A Comparative Analysis of Reconstructed Jaw Musculature and Mechanics of Some Large Theropods," in *Tyrannosaurid Paleobiology*, Parrish, J.M., Molnar, R.E., Currie, P.J., and Koppelhus, E.B. (eds.), Bloomington, IN: Indiana University Press, 2013: 177–193.

Moore, J.A., *Science As a Way of Knowing: The Foundations of Modern Biology*. Cambridge, MA: Harvard University Press, 1993.

Morris, E., *The Ashtray (or the Man Who Denied Reality)*. Chicago: University of Chicago Press, 2018.

Muller, J.Z., *The Tyranny of Metrics*. Princeton: Princeton University Press, 2018.

Murphy, N.L., Carpenter, K., and Trexler, D., "New Evidence for Predation by a Large Tyrannosaurid," in *Tyrannosaurid Paleobiology*, Parrish, J.M., Molnar, R.E., Currie, P.J., and Koppelhus, E.B. (eds.), Bloomington, IN: Indiana University Press, 2013.

Nagel, T., "What Is It Like to Be a Bat?," *The Philosophical Review*, LXXXIII, 4 (October 1974): 435–450.

NASA: Global Climate Change, https://climate.nasa.gov/evidence

Newman, S., Earthweek: A Diary of the Planet, https://earthweek.com

Newton-Smith, W.H., *The Rationality of Science*. Boston: Routledge & Kegan Paul, 1981.

NOAA Climate.gov, "Understanding Climate," https://www.climate.gov/news-features/understanding-climate

Nye, M.J., "Deoxyribonucleic Acid," in *The Oxford Companion to the History of Modern Science*, Heilbron, J.L. (ed.), Oxford: Oxford University Press, 2003: 203–204.

Oreskes, N., *Why Trust Science?*. Princeton: Princeton University Press, 2019.

Oreskes, N. and Conway, E.M., *Merchants of Doubt: How a Handful of Scientists Obscured the Truth from Tobacco Smoke to Climate Change*. New York: Bloomsbury Press, 2011.

Padian, K. and Chiappe, L.M., "Bird Origins," in *Encyclopedia of Dinosaurs*, Currie, P.J. and Padian, K. (eds.), San Diego: Academic Press, 1997.

Parsons, K.M., "The Wrongheaded Dinosaur," *Carnegie Magazine*, (November/December 1997), https://carnegiemuseums.org/magazine-archive/1997/novdec/feat5.htm

Parsons, K.M., *Drawing Out Leviathan: Dinosaurs and the Science Wars*. Bloomington, IN: Indiana University Press, 2001.

Parsons, K.M., "Reviving Pagan Spirituality: A Manifesto," *Religions*, 13: 10; (October 2022). Article is available online at: 10.3390/rel13100942

Parsons, K.M. and Zaballa, R.A., *Bombing the Marshall Islands: A Cold War Tragedy*. Cambridge: Cambridge University Press, 2017.

PBS Nova, "Can We Cool the Planet?," season 47, episode 15.

Pera, M., *Discourses of Science*, Botsford, C. (trans.), Chicago: The University of Chicago Press, 1994.

Pinker, S., *The Blank Slate: The Modern Denial of Human Nature*. New York: Penguin Group, 2002.

Pinker, S., *Enlightenment Now: The Case for Reason, Science, Humanism, and Progress*. New York: Viking, 2018.

Pinnick, C.L., "What's Wrong with the Strong Programme's Case Study of the Hobbes/Boyle Dispute," in *A House Built on Sand: Exposing Postmodernist Myths about Science*, Koertge, N. (ed.), Oxford: Oxford University Press, 1998.

Pinnick, C.L., "Problems with Feminist Epistemology," in *Scientific Inquiry: Readings In the Philosophy of Science*, Klee, R. (ed.), Oxford: Oxford University Press, 1999: 295–305. First published as: "Feminist Epistemology: Implications for the Philosophy of Science," *Philosophy of Science* 61, 1994: 646–57.

Preston, J., *Kuhn's The Structure of Scientific Revolutions: A Reader's Guide*. London: Continuum International Publishing Group, 2008.

Raup, D.M., *The Nemesis Affair*. New York: W.W. Norton, 1986.

Raup, D.M., *Extinction: Bad Genes or Bad Luck?*. New York: W.W. Norton, 1991.

Rhodes, R., *The Making of the Atomic Bomb*. New York: Simon and Schuster, 1986.

Ritchie, S., *Science Fictions: How Fraud, Bias, Negligence, and Hype Undermine the Search for Truth*. New York: Henry Holt and Company, 2020.

Rolin, K., "Tensions among Ideals," in *The Rightful Place of Science: Science, Values, and Democracy*, Richards, T. (ed.), Tempe, AZ: The Consortium for Science, Policy and Democracy, 2021: 37–50.

Romm, J., *Climate Change: What Everyone Needs to Know*. Oxford: Oxford University Press, 2016.

Rorty, R., *Philosophy and the Mirror of Nature*. Princeton: Princeton University Press, 1979.

Rorty, R., "The World Well Lost," in *Consequences of Pragmatism*, Minneapolis: University of Minnesota Press, 1982: 3–18.

Rorty, R., "Phony Science Wars," *The Atlantic*, November, 1999, https://www.theatlantic.com/magazine/archive/1999/11/phony-science-wars/377882/

Rorty, R., "Kuhn," in *A Companion to the Philosophy of Science*, Newton-Smith, W.H. (ed.), Oxford: Blackwell Publishers, 2000: 203–206.

Salmon, W., "Rationality and Objectivity in Science," in *Reality and Rationality*, Dowe, P. and Salmon, M.H. (eds.), Oxford: Oxford University Press, 2005: 93–116.

Searle, J., *Mind, Language, and Society: Philosophy in the Real World*. New York: Basic Books, 1998.

Sepkoski, D. and Ruse, M. (eds.), *The Paleobiological Revolution: Essays on the Growth of Modern Paleontology*. Chicago: Chicago University Press, 2009.

Sepkoski, D. and Ruse, M., "Introduction: Paleontology at the High Table," in *The Paleobiological Revolution: Essays on the Growth of Modern Paleontology*, Sepkoski, D. and Ruse, M. (eds.), Chicago: Chicago University Press, 2009.

Shapin, S. and Schaffer, S., *Leviathan and the Air-Pump*. Princeton, Princeton University Press, 1985.

Shubin, S., *Your Inner Fish: A Journey into the 3.5 Billion-Year History of the Human Body*. New York: Pantheon Books, 2008.

Shubin, S., *Some Assembly Required: Decoding Four Billion Years of Life, from Ancient Fossils to DNA*. New York: Pantheon Books, 2020.

Simpson, G.G., "Recent Advances in Methods of Phylogenetic Inference," in *Phylogeny of the Primates*, Luckett, W.P. and Szalay, F.S. (eds.), New York Plenum Press, 1975: 3–19.

Soh, D., *The End of Gender: Debunking the Myths about Sex and Identity in Our Society*. New York: Simon & Schuster, 2020.

Sokal, A. "Transgressing the Boundaries: Toward a Transformative Hermeneutics of Quantum Gravity," in *The Sokal Hoax: The Sham that Shook the Academy*, Franca, L. (eds.), Lincoln, NE: University of Nebraska Press, 2000: 11–45.

Sokal, A. *Beyond the Hoax: Science, Philosophy, and Culture*. Oxford: Oxford University Press, 2008.

Sokal, A. and Bricmont, J., *Fashionable Nonsense: Postmodern Intellectuals' Abuse of Science*. New York: Picador, USA, 1998.

Thewissen, J.G.M., *The Walking Whales: From Land to Water in Eight Million Years*. Oakland, CA: University of California Press, 2014.

Thornton, B.S., *Plagues of the Mind: The New Epidemic of False Knowledge*. Wilmington, DE: ISI Books, 1999.

Toulmin, S., *Human Understanding: The Collective Use and Evolution of Concepts*. Princeton: Princeton University Press, 1972.

Wagner, P. and Fair, F., *Education for Knowing*. New York: Rowman & Littlefield, 2020.

Watson, J.D., *The Double Helix*, The Annotated and Illustrated Edition, Gann, A. and Witkowski, J. (eds.), New York: Simon & Schuster, 2012.

Weart, S.R., *The Discovery of Global Warming*, Revised and Expanded Edition. Cambridge, MA: Harvard University Press, 2008.

Weiner, J., *The Beak of the Finch: A Story of Evolution in Our Time*. New York: Vintage Books, 1995.

Weinert, F., *Copernicus, Darwin, and Freud: Revolutions in the History and Philosophy of Science*. Chichester, West Sussex, UK: Wiley-Blackwell, 2009.

Wootton, D., *The Invention of Science: A New History of the Scientific Revolution*. New York: HarperCollins, 2015.

Yohe, G., "A New Book Manages to Get Climate Science Badly Wrong," *Scientific American*, May 13, 2021: https://www.scientificamerican.com/article/a-new-book-manages-to-get-climate-science-badly-wrong/

Zammito, J.H., *A Nice Derangement of Epistemes: Post-Positivism in the Study of Science from Quine to Latour*. Chicago: The University of Chicago Press, 2004.

Index

aerosols 192
Angstrom, Knut 188, 189
anthropogenic Climate Change (ACC) 16, 172–175, 182–197
antirealism 62, 63, 97–100
applied ethics 127, 128
Archaeopteryx lithographica 156
Aristotle 4, 18, 52, 71, 94, 95, 100
Arrhenius, Svante 186–190
"ashtray incident" 61, 62
atom bomb test *see* Trinity Test

background knowledge 36
Bakker, Robert 142, 143
Bayes' Theorem 90, 91
Bell Curve, The 117
Bernstein, Richard 87–90, 123
bias in publication 205–208
Bird, Alexander 98, 99
bird evolution 14, 155–159
Blondlot, Rene Prosper 45–47, 101, 204
Boghossian, Paul 23–25, 40
Boyle, Robert 49–52, 56–59
Bricmont, Jean 66
Brown, Matthew J. 130, 131, 136
"business Model" of universities 17, 18, 209, 210

Cepheid variables 53–55
Chalmers, A.F. 67, 73
cladistics 156–158

climate sensitivity 193, 194
Climatology 16, 17, 183–197
consilience 16, 182, 196
"conversion" 15, 68, 160–163, 166, 167
Copernican Revolution 169
"creation science" 40
Cruz, Ted 198–201
Curie, Marie 105

Darwin, Charles 9, 62, 78–82, 103, 117, 176, 177, 210, 217
Death of the Ball Turret Gunner 6
Deinonychus antirrhopus 156
Dessler, Andrew E. 190–193
dinosaurs 8, 14, 15, 29, 40, 47–49, 139–159
discovery 83, 84
diversity 42, 112, 113, 129
DNA 43, 84, 122–123
Douglas, Heather 14, 115–130, 132, 136, 218
Duhem, Pierre 31

Einstein, Albert 9, 210
embryology 16, 179, 180
encephalization quotient 144
evolution 15; confirmation of 175–183; five theories of (Mayr) 175, 176
experiment 31, 32, 49, 50, 59
Extinction: gradual vs. Sudden 164, 165

extinction of the Dinosaurs (K/Pg extinction event) 14, 15, 159–162, 164

false positives and false negatives 124
Fastovsky, David E. 150–155
feathered dinosaurs 155
feedback loops 172, 187, 193, 194
Finite element analysis (FEA) 147, 152
Fish, Stanley 22, 23
feminism 14, 21, 105–115, 135; as ideology 113, 114, 135; epistemic value of 130–136; gender feminism 111, 133–135
Feynman, Richard 9

Galileo, Galilei 69–74, 98, 103
gestalt switches 64, 68, 69, 121, 161, 167
Gimbutas, Marija 133
"global" vs. "local" challenges to science 15, 171, 172
"Goodheart's Law" 212
Gould, Stephen Jay 43
"greenhouse gases" 184
Gribbin, John 74
Gross, Paul R. 21, 58, 59, 132

h-index 211
Haack, Susan 115
Hacking, Ian 26, 63, 65
Harding, Sandra 14, 63, 105–115, 130
Hobbes, Thomas 49–52, 56–59
holism thesis (HT) 11, 32, 33, 39, 40
Holmes, Sherlock 176, 177, 218, 219
homologies 78–82
How to Lie With Statistics 201
Huxley, Thomas Henry 156

impact hypothesis, *see* extinction of the dinosaurs
incommensurability 13, 15, 65, 76–82, 93–97, 160, 168
indeterminacy of translation (Quine) 215, 216
inference to the best explanation 16, 176, 177, 218, 219
inferential gap (Douglas) 119, 120, 129
irrationality of theory choice 64–66

Johnson, Phillip 103, 130

Khalidi, Muhammad Ali 93, 94
Klein, Ellen R. 110–112
Koonin, Steven E. 174, 185
Krauss, Lawrence 187–190
Kuhn, Thomas 12, 13, 18, 61–68, 83, 84, 85, 100, 112, 139, 160, 161, 168, 169, 213, 216, 217; Kuhn and rationality 68–76; Kuhn and incommensurability 76–82, 93–97; post-*Structure* writings 86–93; Kuhn's antirealism 97–100

Latour, Bruno 12, 44–49
Laudan, Larry 38, 39
Lavoisier, Antoine 33
Leavitt, Henrietta Swan 53–56, 113
Levitt, Norman 21, 58, 59, 132
Lyell, Charles 163

Maudlin, Tim 74, 75
Mayr, Ernst 175–177, 185
McIntyre, Lee 198, 199
metrics 18, 210–212
Morris, Errol 61–63, 77
Mother Goddess, see Gimbutas, Maria
Muller, Jerry Z. 18, 211

N-rays *see* Blondlot, Rene Prosper
Nagel, Thomas 39, 40
natural selection 180, 181
Nemesis hypothesis (Raup) 162, 167, 168

objections to anthropogenic climate change 195, 196
objectivity 14, 29, 105, 106, 108–115
observation 13, 70–75
Oreskes, Naomi 112, 113, 174
orwell, George 48
Owen, Richard 49, 62, 78–82, 102, 155

p-hacking 206–208
paleobiology 141–155
paleoclimatology 183
paradigms (Kuhn) 64, 77, 83, 142, 165, 166, 168
perverse incentives 208, 209, 212
phases of Venus 96
phlogiston 33
"phony wars" (Rorty) 26–28
phronesis see practical wisdom
Pinker, Steven 11, 13, 111, 134
Pinnick, Cassandra 56–59, 109, 110
planet, definition of *see* Pluto
Pluto 82
political rectitude 30, 105–115
Population Bomb, The 184
postmodernism 19, 21, 24
practical wisdom 18, 88, 89
Pre-registration of trials 213
pressure to publish 209, 210
pre-test criteria 35–37
pseudo-skeptics *see* skepticism
Putnam, Hilary 77, 78

Quine, W.V.O. 11, 31, 32, 215

radiative forcing 190–193
Raup, David 15, 161–167, 169

Rayfield, Emily 147, 152, 153
regulative assumptions 35, 36
Relativism 11, 12, 13, 19, 20, 24, 25, 27, 30–42, 61, 65, 66, 68
replication 204, 205
Ritchie, Stuart 17, 115, 204–214
Robinson Crusoe 38
Rolin, Kristina 131, 132
Rorty, Richard 18, 26–28, 47, 97, 98

"salami slicing" 209, 210, 214
Salmon, Wesley 90–93
Schaffer, Simon 12, 49–52, 56–60
Shapin, Stephen 12, 49–52, 56–60
Shubin, Neil 178–180
science Wars 10, 15, 26–28
scientific methods 12, 34, 35, 49–60, 150–155
scientific rationalism 28, 29, 38, 67, 167
Scientific Revolution, The 11
scientific revolutions 14, 64, 67, 169
skepticism and pseudo-skepticism 12, 16, 17, 172, 197–202
social constructivism 12, 15, 20, 30, 43–60
Social Text 21, 22, 24, 25
Soh, Debra 111, 134, 135
Sokal, Alan 10, 22, 66
Sokal hoax 10, 22–25
"stamp collecting" (Rutherford) 142
standpoint epistemology (Sandra Harding) 107–109, 112

Teller, Edward 121
"Texas sharpshooter" 207, 213
theoretical virtues 36, 37
Thornton, Bruce 133
Tiktaalik 178, 179

Toles, Tom 118
Tolstoy, Leo 160, 161, 168
Toulmin, Stephen 169
translation 93–96
Trinity Test 6, 7, 127
Tyrannosaurus rex 143; As predator 148–150; bite 145–147; brain and senses 144, 145, 149; hunting technique 149, 150; limbs 148–149
Tuskegee Experiment 125

Ulam, Stanislaw 121
underdetermination thesis (UT) 11, 33–41
Uniformitarianism 163, 164
United Nations' Intergovernmental Panel on Climate Change 129, 173

"value free" science 136, 137
values, epistemic vs. non-epistemic 104

Wagner, Paul 216, 217
"ways of knowing" 24, 28
Weinert, Friedel 35
Whewell, William 163
women in science 105, 106
Wooton, David 4, 11, 59
"world changes" (Kuhn) 64, 69, 98
whale evolution 179
Whig history 61, 62, 100–102
"wrongheaded" dinosaur 43

"year without a summer" 192
Yohe, Gary 174, 185

Zammito, John H. 56–59, 215

For Product Safety Concerns and Information please contact our EU representative GPSR@taylorandfrancis.com
Taylor & Francis Verlag GmbH, Kaufingerstraße 24, 80331 München, Germany

www.ingramcontent.com/pod-product-compliance
Lightning Source LLC
Chambersburg PA
CBHW071817230426
43670CB00013B/2482